LIFE SCIENCE APPLICATIONS FOR PHYSICS

A MANUAL to accompany
Principles of Physics, second edition and
Physics for Scientists and Engineers, fourth edition
by Ray Serway

Compiled by
Jerry Faughn
Eastern Kentucky University

THOMSON
BROOKS/COLE

Australia • Canada • Mexico • Singapore • Spain • United Kingdom • United States

Printed in the United States of America

3 4 5 6 7 05 04 03 02

0-03-020662-6

For more information about our products, contact us at:
Thomson Learning Academic Resource Center
1-800-423-0563

For permission to use material from this text, contact us by:
Phone: 1-800-730-2214
Fax: 1-800-731-2215
Web: www.thomsonrights.com

Asia
Thomson Learning
5 Shenton Way #01-01
UIC Building
Singapore 068808

Australia
Nelson Thomson Learning
102 Dodds Street
South Street
South Melbourne, Victoria 3205
Australia

Canada
Nelson Thomson Learning
1120 Birchmount Road
Toronto, Ontario M1K 5G4
Canada

Europe/Middle East/South Africa
Thomson Learning
High Holborn House
50/51 Bedford Row
London WC1R 4LR
United Kingdom

Latin America
Thomson Learning
Seneca, 53
Colonia Polanco
11560 Mexico D.F.
Mexico

Spain
Paraninfo Thomson Learning
Calle/Magallanes, 25
28015 Madrid, Spain

PREFACE

The material in this booklet has been compiled as a supplement to *Physics for Scientists and Engineers, 4e,* and *Principles of Physics, 2e*, by Raymond A Serway. The purpose of this material is to provide readings, examples, and problems having applications to the life sciences. The topics covered will be of special interest to students who are pursuing a biology major or a pre-professional program in medicine, pharmacy, etc. A table of contents with cross-references to specific chapters in the two texts is included for convenience in making class assignments.

Most of the material in this manual is taken from the various editions of Serway and Faughn's *College Physics*. The sections concerning Poiseuille's Law, Basal Metabolic Rates, and The Physics of Vision are taken from *Physics for Biology and Pre-med Students* by Leonard H. Greenburg (W.B. Saunders Co., 1975).

Table of Contents

Corresponding chapters in PSE (*Physics for Scientists and Engineers, 4e*) and POP (*Principles of Physics, 2e*) are listed below each section heading.

Section 1
MECHANICS OF THE HUMAN BODY

Mechanics is that portion of physics that deals with motion and the forces that cause motion. One of the greatest breakthroughs that has led to our present understanding of mechanics occurred when Newton presented his three laws of motion in the seventeenth century. So important are these laws, that much of the first semester of an introductory physics course is devoted to learning how to apply them to specific physical situations. We are concerned here with applications of physics to the life sciences; thus we shall begin our study with some worked out examples that show how Newton's first, second, and third laws can be used in situations involving the human body. As we shall see, a medical doctor placing a patient in traction, an orthodontist straightening a tooth, or you, when you chew your food, make use of these fundamental laws of physics.

PROBLEMS
MECHANICS OF THE HUMAN BODY

1. Figure 1.1 below illustrates the difference in proportions of the male and female anatomy. The displacement $\mathbf{d}_{1m}$ and $\mathbf{d}_{1f}$ from the bottom of the feet to the navel have magnitudes of 104 cm and 84.0 cm, respectively. The displacements $\mathbf{d}_{2m}$ and $\mathbf{d}_{2f}$ have magnitudes of 50.0 cm and 43.0 cm, respectively. (a) Find the vector sum of the displacements $\mathbf{d}_1$ and $\mathbf{d}_2$ in each case. (b) The male figure is 180 cm tall, the female 168 cm. Normalize the displacements of each figure to a common height of 200 cm and reform the vector sums as in part (a). Then find the vector difference between the two sums.

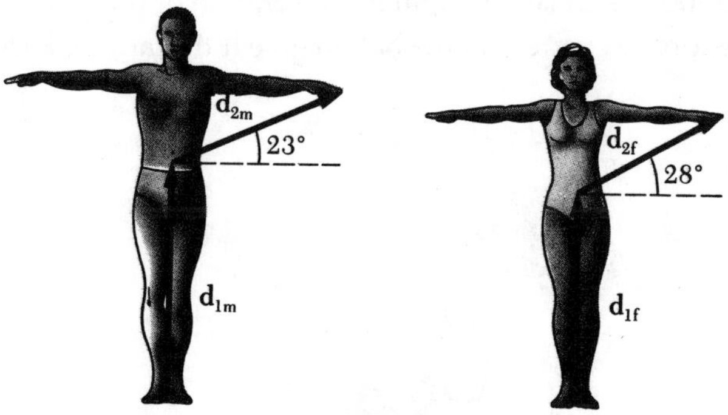

Figure 1.1

2. The leg in the cast shown in Figure 1.2 below weighs 220 N ($\mathbf{w}_1$). Determine the weight $\mathbf{w}_2$ and the angle α in order that there be no force exerted on the hip joint *J* by the leg in the cast.

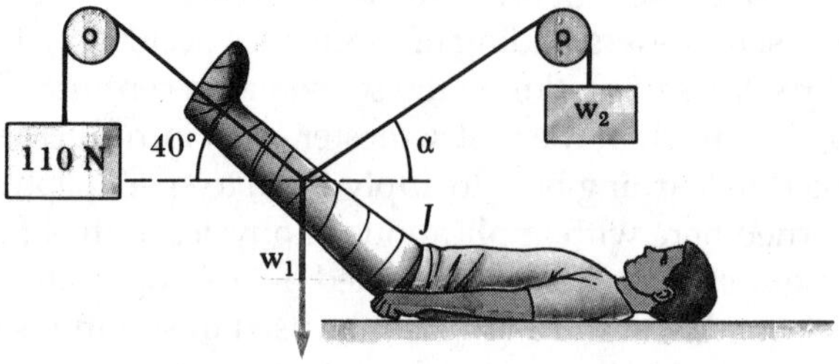

Figure 1.2

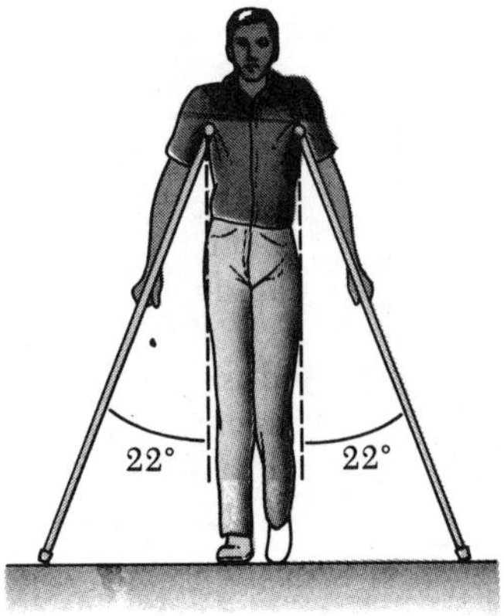

Figure 1.3

3. A person standing on crutches as shown in Figure 1.3 weighs 170 lb. The crutches each make an angle of 22° with the vertical (as seen from the front). Half of the person's weight is supported by the crutches. The other half is supported by the vertical forces of the ground on the feet. Assuming the person is at rest and the force of the ground on the crutches acts along the crutches, determine (a) the coefficient of friction of the crutches on the ground and (b) the magnitude of the compression force supported by each crutch.

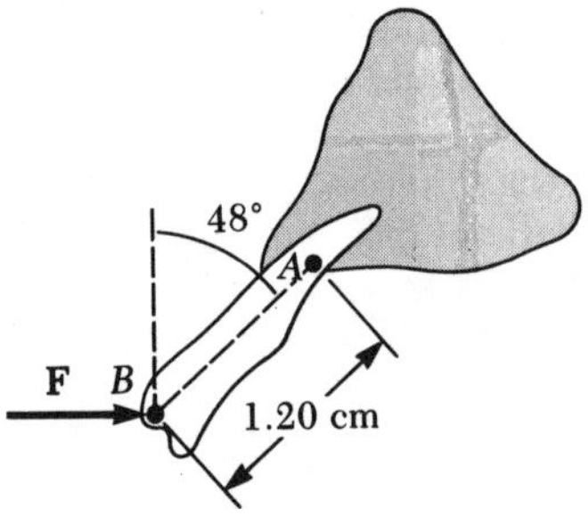

Figure 1.4

4. A steel band exerts a horizontal force of 80.0 N on a tooth at the point B in Figure 1.4. What is the torque on the root of the tooth about point A?

5. A baseball player loosening up his arm before a game tosses a 0.15 kg baseball using only the rotation of his forearm to accelerate the ball, as shown in Figure 1.5. The ball starts from rest and is released with a speed of 15 m/s in 0.3 s. (a) Find the constant angular acceleration of the arm and the ball. (b) Find the torque exerted on the ball to give it this angular acceleration.

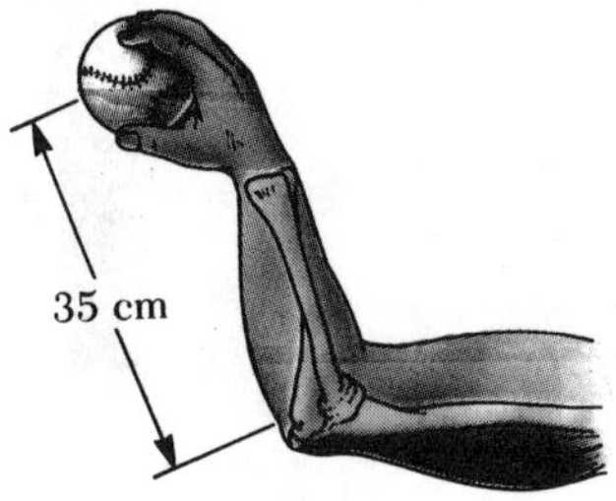

Figure 1.5

6. The muscle used for chewing, the masseter, is one of the strongest in the human body. It is attached to the mandible, or lower jawbone, as shown in Figure 1.6 (a) below. The jawbone is pivoted about a socket just in front of the auditory canal. The forces acting on the jawbone are equivalent to those acting on a curved bar as shown in part (b) of Figure 1.6. The force **C** is the force exerted against the jawbone by the food being chewed, **T** is the tension in the masseter, and **R** is the force exerted on the mandible by the socket. If you bite down on a piece of steak with a force of 50 N, find the forces **T** and **R.**

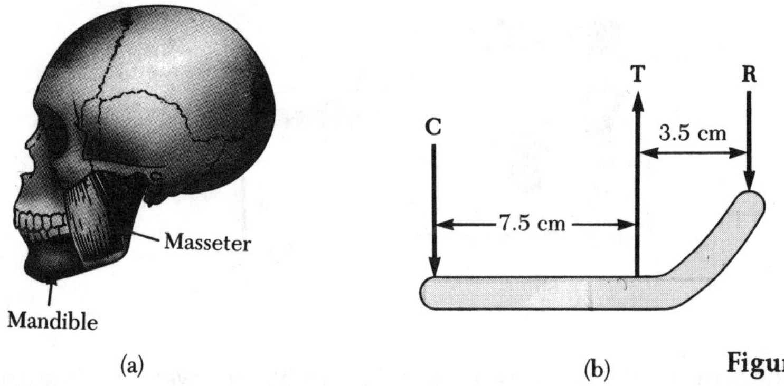

Figure 1.6

7. The large quadriceps muscle in the upper leg terminates at its lower end in a tendon attached to the upper end of the tibia as shown in Figure 1.7 (a) below. The forces on the lower leg when the leg is extended are modeled as in part (b) of Figure 1.7, where **T** is the tension in the tendon, **C** is the weight of the lower leg, and **F** is the weight of the foot. Find the tension T when the tendon is at an angle of 25° with the tibia, assuming $C = 30$ N, $F = 12.5$ N, and the leg is extended at an angle of 40° with respect to the vertical ($\theta = 40°$). Assume the center of gravity of the lower leg is at its center, and that the tendon attaches to the lower leg at a point one fifth of the way down the leg.

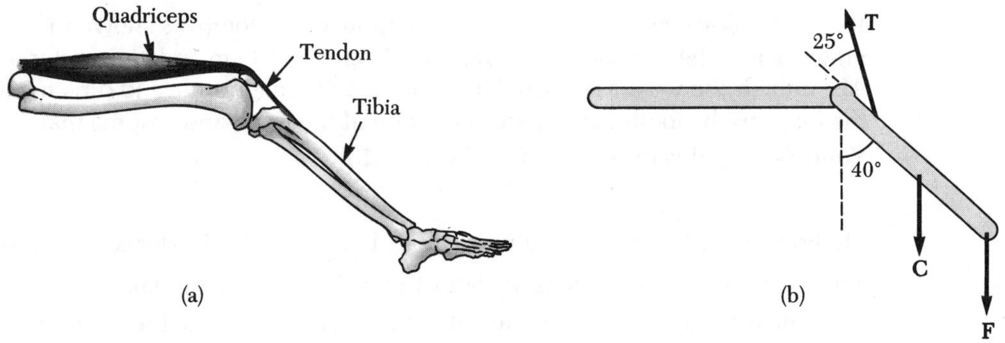

Figure 1.7

8. A person bends over and lifts a 200 N weight as shown in Figure 1.8 (a) below, with the back in the horizontal position. The back muscle, attached at a point two thirds up the spine, maintains the position of the back, where the angle between the spine and this muscle is 12°. Using the mechanical model shown in Figure 1.8 (b) and taking the weight of the upper body to be 350 N, find the tension in the back muscle and the compressional force in the spine.

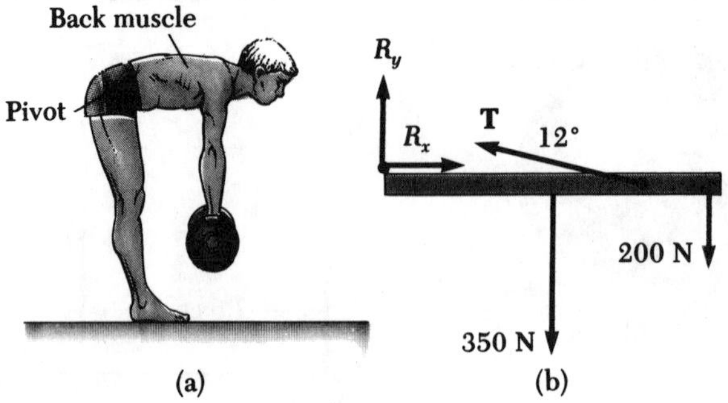

Figure 1.8

9. The arm shown in Figure 1.9 weighs 41.5 N. The weight of the arm acts through point A. Determine the magnitudes of the tension force $\mathbf{F_t}$ in the deltoid muscle and the force $\mathbf{F_s}$ of the shoulder on the humerus (upper arm bone) to hold the arm in the position shown.

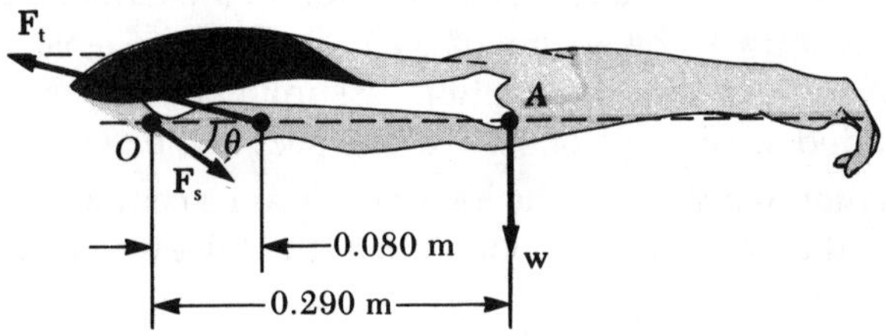

Figure 1.9

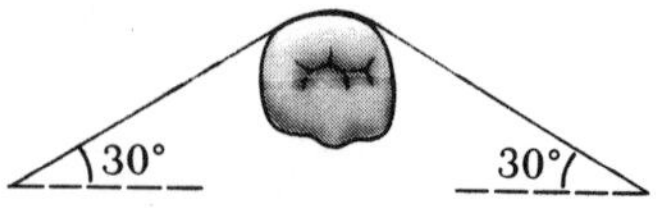

Figure 1.10

10. A stainless-steel orthodontic wire is applied to a tooth, as shown in Figure 1.10. The wire has an unstretched length of 3.1 cm and a diameter of 0.22 mm. If the wire is stretched 0.1 mm, find the magnitude and direction of the force on the tooth. Disregard the width of the tooth and assume that Young's modulus for stainless steel = 18×10^{10} Pa.

11. Bone has a Young's modulus of about 1.45×10^9 Pa. Under compression, a bone can withstand a stress of about 1.60×10^8 Pa before breaking. Estimate the length of your femur (thigh bone) and calculate the amount this bone can be compressed before breaking.

12. Figure 1.11 below shows a crude model of an insect wing. The mass *m* represents the entire mass of the wing, which pivots about the fulcrum F. The spring represents the surrounding connective tissue. Motion of the wing corresponds to vibration of the spring. Suppose the mass of the wing is 0.3 g and the effective spring constant of the tissue is 4.7×10^{-4} N/m. If the mass moves up and down a distance of 2.0 mm from its position of equilibrium, what is the maximum speed of the outer tip of the wing?

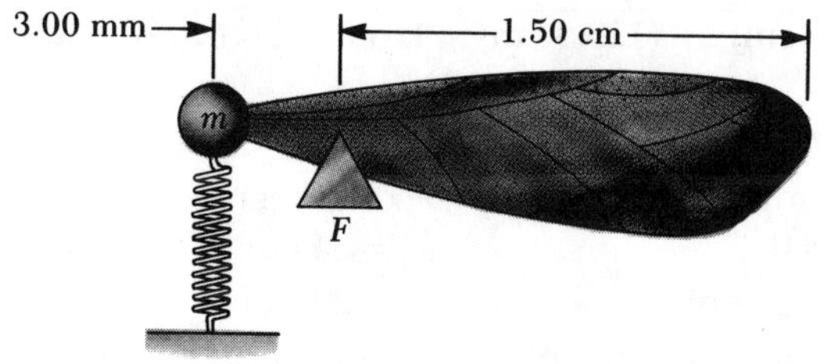

Figure 1.11

SOLUTIONS

MECHANICS OF THE HUMAN BODY

1. (a) For the male anatomy, we have

x-component	*y* -component
$d_{1x} = 0$	$d_{1y} = 104$ cm
$d_{2x} = 46$ cm	$d_{2y} = 19.5$ cm

For a resultant component of $d_x = 46$ cm, and a resultant *y* component $d_y =$ 123.5 cm.

So, $d = \sqrt{(46\text{ cm})^2 + (123.5\text{ cm})^2}$; $\tan\theta = 2.68$, and $\theta = 69.6°$.

For the female anatomy, we have

x-component	*y* component
$d_{1x} = 0$	$d_{1y} = 84$ cm
$d_{2x} = 38$ cm	$d_{2y} = 20.2$ cm

For a resultant x component of $d_x = 38$ cm, and a resultant y component $d_y =$ 104.2 cm.

So, $d = \sqrt{(38\text{ cm})^2 + (1104.2\text{ cm})^2}$; $\tan\theta = 2.74$, and $\theta = 70°$.

(b) To normalize, multiply all distances by the appropriate scale factors which are: $s_m = \dfrac{200\text{ cm}}{180\text{ cm}} = 1.111$ and $s_f = \dfrac{200\text{ cm}}{168\text{ cm}} = 1.190$

Multiplying all distances by these scale factors and recomputing the sums, yields:

$d_m' = 146.4$ cm and the angle is 69.6°
$d_f' = 132.0$ cm and the angle is 70°

To compute the vector difference

$\Delta \mathbf{d} = \mathbf{d_m}' - \mathbf{d_f}' = \mathbf{d_m}' + (-\mathbf{d_f}')$, we have

x-component	*y* -component
$d_{mx}' = 51.0$ cm	$d_{my}' = 137.2$ cm
$-d_{fx}' = -45.1$ cm	$-d_{fy}' = -124.0$ cm

For a resultant *x* component of 5.9 cm, and a resultant *y* component of 13.2 cm
The Pythagorean theorem yields, $\Delta d = 14.5$ cm

and $\tan\theta = \dfrac{13.2\text{ cm}}{5.9\text{ cm}} = 2.24$, from which $\theta = 65.9°$.

2. From $\Sigma F_x = 0$, we have

$$W_2\cos\alpha - (110\text{ N})\cos 40° = 0 \quad (1)$$

From $\Sigma F_y = 0$, we have

$$W_2\sin\alpha + (110\text{ N})\sin 40° - 220\text{ N} = 0 \quad (2)$$

Dividing (2) by (1) yields

$$\tan\alpha = \frac{149.3}{84.26} = 1.772$$

$$\alpha = 60.55°$$

Then, from either (1) or (2), $W_2 = 171.4$ N.

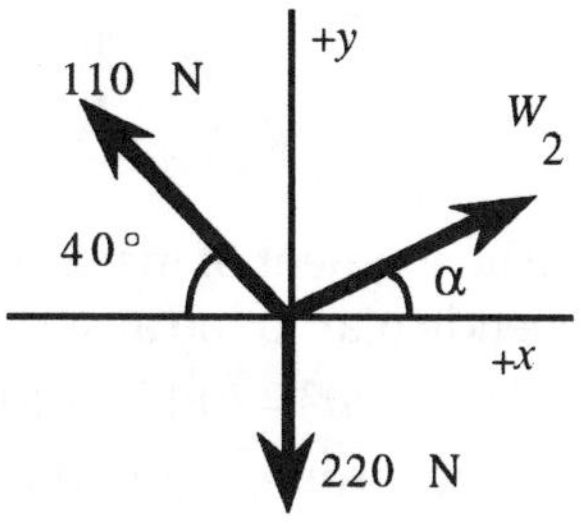

Figure 1.12

3. From the free-body diagram of the person, find the compression force acting along the line of each crutch.

$\Sigma F_x = 0$ gives

$F_1 \sin 22° - F_2 \sin 22° = 0$

or $\quad F_1 = F_2 = F$

Then $\Sigma F_y = 0$ gives

$2F\cos 22° - 85\text{ lb} = 0$

From which, $F = 45.8$ lb

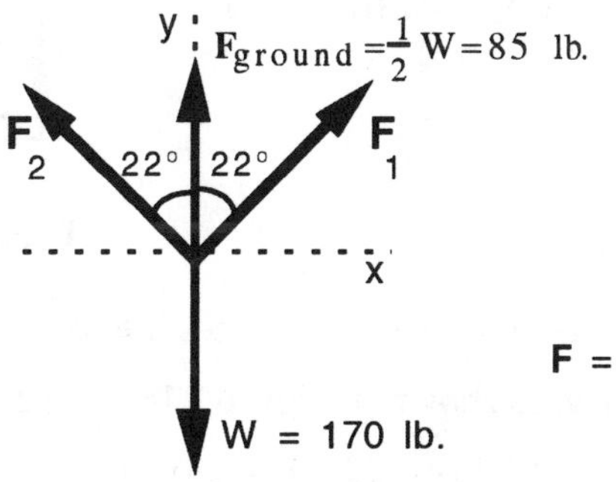

Free-Body Diagram of Person

Figure 1.13

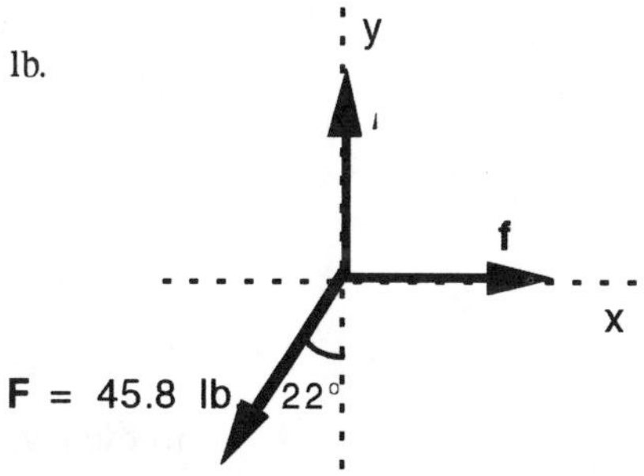

Free-Body Diagram of Crutch Tip

Figure 1.14

Now consider a crutch tip as shown. $\Sigma F_x = 0$ becomes

$f - (45.8\text{ lb})\sin 22° = 0$ or

$f = 17.2$ lb

$\Sigma F_y = 0$ becomes

$N - (45.8\text{ lb})\cos 22° = 0$ or $\quad N = 42.5$ lb

Assuming it is on the verge of slipping, the minimum possible coefficient of friction is

$$\mu_s = \frac{f}{N} = \frac{17.2\text{ lb}}{42.5\text{ lb}} = 0.404$$

4. The lever arm is $d = (1.2 \times 10^{-2}\text{ m})\cos 48° = 8.03 \times 10^{-3}$ m, and the torque is

$$\tau = Fd = (80.0\text{ N})(8.03 \times 10^{-3}\text{ m}) = 0.642\text{ N}\cdot\text{m counterclockwise}$$

5. **(a)** During its acceleration, the ball moves through an arc of a circle having a radius of 0.35 m. We can determine the angular acceleration using

$\omega = \omega_0 + \alpha t$. Since $\omega_0 = 0$, $\omega = \alpha t$, so $\alpha = \dfrac{\omega}{t}$

We also know that v = rω, so that we get

$$\alpha = \frac{\omega}{t} = \frac{v}{r\,t} = \frac{15 \text{ m/s}}{(0.35 \text{ m})(0.3 \text{ s})} = 143 \text{ rad/s}^2$$

(b) The moment of inertia of the ball about an axis through the elbow and perpendicular to the arm is

$$I = mr^2 = (0.15 \text{ kg})(0.35 \text{ m})^2 = 1.84 \times 10^{-2} \text{ kg m}^2$$

Thus, the torque required is

$$\tau = I\alpha = (1.84 \times 10^{-2} \text{ kg m}^2)(143 \text{ rad/s}^2) = 2.63 \text{ N} \cdot \text{m}$$

6. Choosing the pivot point at the point *O* shown, $\Sigma\tau = 0$ becomes

$(50 \text{ N})(7.5 \text{ cm}) + T(0) - R(3.5 \text{ cm}) = 0$

Thus, $R = 107$ N

Now, apply $\Sigma F_y = 0$.

$-50 \text{ N} + T - 107 \text{ N} = 0$

and $T = 157$ N

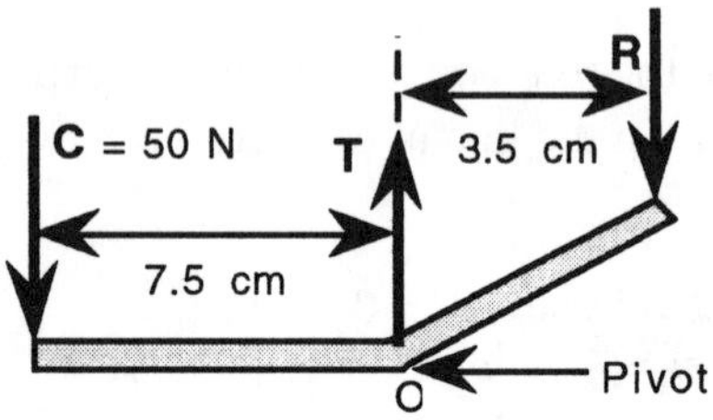

Figure 1.15

7. Let us first resolve all forces into components parallel and perpendicular to the leg, as shown in Figure 1.16. Use $\Sigma\tau = 0$ about the pivot indicated.

$$T_y(d/5) - C_y(d/2) - F_y(d) = 0$$

d is the length of the lower leg.

$C_y = C \sin V = (30 \text{ N})\sin 40° = 19.3$ N, and

$F_y = F\sin V = (12.5 \text{ N})\sin 40° = 8.03$ N.

Thus, $T_y = 88.5$ N, but $T_y = T \sin 25°$

So, $T = 209$ N

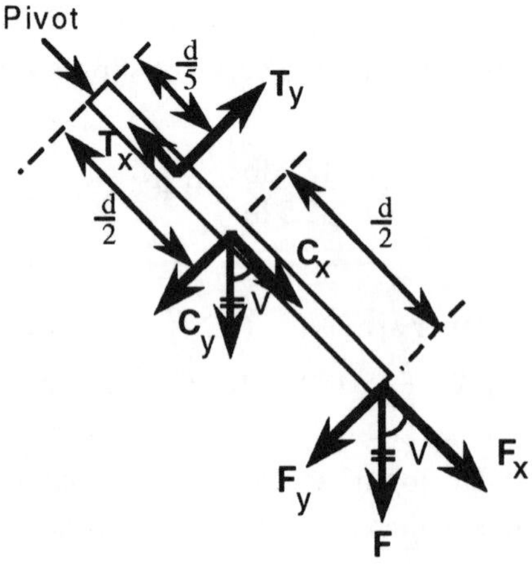

Figure 1.16

8. The free body diagram is shown at the right.
$\Sigma\tau = 0$, yields

$$-\frac{L}{2}(350\text{ N}) + T\sin 12^\circ\left(\frac{2L}{3}\right) - (200\text{ N})L = 0$$

From which, $T = 2705$ N
Compression force along the spine = R_x, and we find this from $\Sigma F_x = 0$, which gives

$$R_x = T_x = T\cos 12^\circ = 2646\text{ N}$$

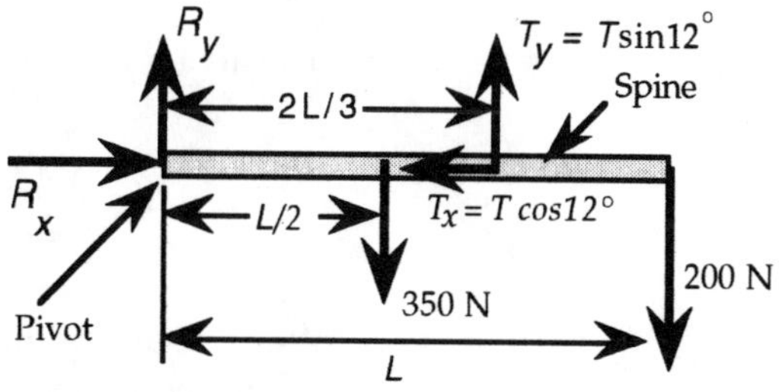

Figure 1.17

9.

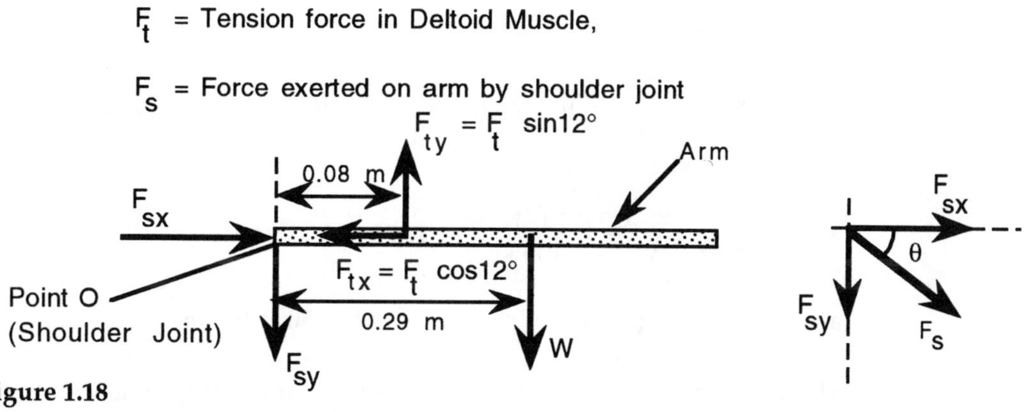

Figure 1.18

$\Sigma\tau_O = F_t\sin 12^\circ(0.08\text{ m}) - (41.5\text{ N})(0.29\text{ m}) = 0$
From which $F_t = 724$ N (Tension in deltoid muscle)

$\Sigma F_y = 0$ gives $\quad -F_{sy} + F_t\sin 12^\circ - 41.5\text{ N} = 0$

yielding, $\quad F_{sy} = 109$ N

$\Sigma F_x = 0$ gives $\quad F_{sx} = F_t\cos 12^\circ$ and $\quad F_{sx} = 709$ N

Therefore, $F_s = \sqrt{(F_{sx})^2 + (Fs_y)^2} = 717\text{N}$

and $\quad \tan\theta = \dfrac{F_{sy}}{F_{sx}} = 0.1539 \quad$ and $\theta = 8.75^\circ$

10. Let us find the tension to stretch the wire by 0.1 mm.

The area = A = $\frac{\pi d^2}{4}$ = 3.80×10^{-8} m^2. Thus, the force is

$$\mathbf{F} = \frac{YA\Delta L}{L_O} = \frac{(18 \times 10^{10}\text{Pa})(3.80 \times 10^{-8}\text{m}^2(10^{-4}\text{ m})}{3.1 \times 10^{-2}\text{ m}} = 22.1/\text{N}$$

We have an equilibrium situation, so

$\Sigma F_x = 0$ becomes $F\cos 30° - F\cos 30° = 0$

$\Sigma F_y = 0$ becomes $2F\sin 30° = 2(22.1\text{ N})\sin 30° = 22.1$ N and will be directed down the page in the figure.

11. We assume a length for the femur of 0.5 m. The amount of compression ΔL is given by,

$$\Delta L = \frac{L\,(\text{stress})}{Y} = \frac{(5 \times 10^{-1}\text{ m})(160 \times 10^{6}\text{ Pa})(10^{-4}\text{ m})}{14.5 \times 10^{9}\text{ Pa}} = 5.52 \times 10^{-3}\text{ m} = 5.5\text{ mm}$$

12. $\omega = \sqrt{\frac{k}{m}} = \sqrt{\frac{4.7 \times 10^{-4}\text{ N/m}}{3 \times 10^{-4}\text{ kg}}} = 1.252\text{ rad/s}$

and $v_{max} = \omega A = (1.252\text{ rad/s})(2 \times 10^{-3}\text{ m}) = 2.50 \times 10^{-3}\text{ m/s} = 2.50\text{ mm/s}$

This is the maximum velocity of the wingtip 3 mm from the fulcrum. To get the maximum velocity of the outer tip of the wing, treat the wing as a rigid body rotating about the fulcrum. All parts have the same angular velocity, so

$\frac{v}{r} = \omega$ leads to $\frac{v_{\text{far tip}}}{r_{\text{far tip}}} = \frac{v_{\text{near tip}}}{r_{\text{near tip}}}$

or $v_{\text{far tip}} = (15\text{ mm})\frac{2.50\text{ mm/s}}{3.0\text{ mm}} = 12.5\text{ mm/s} = 1.25\text{ cm/s}$

Box 1.1 FRICTION IN HUMAN JOINTS

The beautiful engineering design that nature has bestowed on the human body is well illustrated in the way proper forces of friction are maintained in the joints of the body. Consider the difficulties that arise with the lubrication of human joints. If there were excessive friction between two adjacent bones, we would not be able to run or complete any other action requiring the rapid movement of joints. On the other hand, if there were not enough friction in the joints, we would find it very difficult to stand upright, for example, because our knees would always be slipping out of alignment.

Figure 1 shows the structure of a typical joint in the human body. The joint is enclosed in a sac containing a lubricating liquid called synovial fluid. The ends of the bones are covered with cartilage, which is a sponge-like material. When the bones are at rest with respect to each other, the cartilage absorbs much of the synovial fluid. This results in an increase in friction between the bones, which enables a person to, for instance, stand upright with ease. When a person runs, increased pressure causes the synovial fluid to be squeezed out of the cartilage and into the joint. This mechanism lubricates the joint, ensuring freedom of movement. The coefficient of friction between bones lubricated with synovial fluid has been determined experimentally as approximately 0.003.

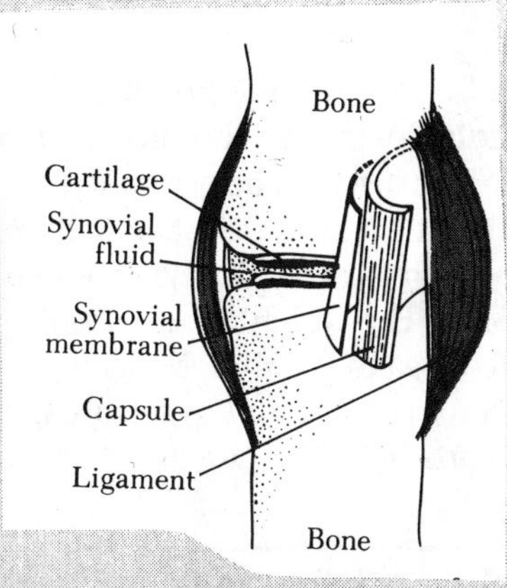

Figure 1
The structure of a typical joint in the human body

Section 2
FLUID MECHANICS AND BIOLOGY

2.1 Surface Tension and Capillarity

Surface Tension

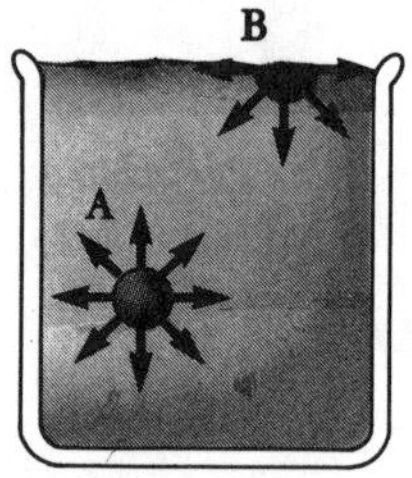

Figure 2.1
The net force on a molecule at A is zero because such a molecule is completely surrounded by other molecules. The net force on a surface molecule at B is downward because it is not completely surrounded by other molecules.

If you look closely at a dewdrop sparkling in the morning sunlight, you will find that the drop is spherical. The drop takes this shape because of a property of liquid surfaces called **surface tension**. In order to understand the origin of surface tension, consider a molecule at point A in a container of water, as in Figure 2.1. Although nearby molecules exert forces on this molecule, the net force on it is zero because it is completely surrounded by other molecules and hence is attracted equally in all directions. The molecule at B, however, is not attracted equally in all directions. Since there are no molecules above it to exert upward forces, the molecule is pulled toward the interior of the liquid. The contraction at the surface of the liquid ceases when the inward pull exerted on the surface molecules is balanced by the outward repulsive forces that arise from collisions with molecules in the interior of the liquid. *The net effect of this pull on all the surface molecules is to make the surface of the liquid contract and consequently to make the surface area of the liquid as small as possible.* Drops of water take on a spherical shape because a sphere has the smallest surface area for a given volume.

If you place a sewing needle very carefully on the surface of a bowl of water, you will find that the needle floats even though the density of steel is about eight times that of water. This also can be explained by surface tension. A close examination of the needle shows that it actually rests in a depression in the liquid surface, as shown in Figure 2.2. The water surface acts like an elastic membrane under tension. The weight of the needle produces a depression, thus increasing the surface area of the film. Molecular forces now act at all points along the depression in an attempt to restore the surface to its original horizontal position. The vertical components of these forces act to balance w, the weight of the needle.

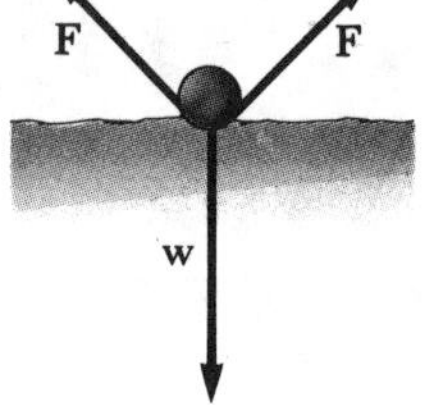

Figure 2.2
End view of a needle resting on the surface of water. The components of surface tension balance the weight force.

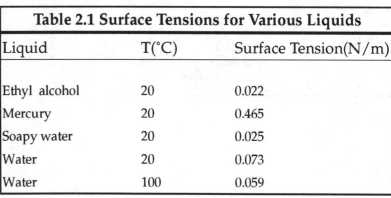

Table 2.1 Surface Tensions for Various Liquids

Liquid	T(°C)	Surface Tension(N/m)
Ethyl alcohol	20	0.022
Mercury	20	0.465
Soapy water	20	0.025
Water	20	0.073
Water	100	0.059

The surface tension, γ, in a film of liquid is defined as the ratio of the magnitude of the surface tension force, F, to the length along which the force acts:

$$\gamma = \frac{F}{L} \tag{2.1}$$

The SI units of surface tension are newtons per meter, and values for a few representative materials are given in Table 2.1

The concept of surface tension can be thought of as the energy content of the fluid at its surface per unit surface area. To see that this is reasonable, we can manipulate the units of surface tension as

$$[\gamma] = \frac{\mathrm{N}}{\mathrm{m}} = \frac{\mathrm{N} \cdot \mathrm{m}}{\mathrm{m}^2} = \frac{\mathrm{J}}{\mathrm{m}^2}$$

In general, *any equilibrium configuration of an object is one in which the energy is a minimum*. Consequently, a fluid will take on a shape such that its surface area is as small as possible. For a given volume, a spherical shape is the one that has the smallest surface area. Therefore, a drop of water takes on a spherical shape.

An apparatus used to measure the surface tension of liquids is shown in Figure 2.3. A circular wire with a circumference ℓ is lifted from a body of liquid. The surface film clings to the inside and outside edges of the wire. This tends to hold back the wire, causing the spring to stretch. If the spring is calibrated, one can measure the force required to overcome the surface tension of the liquid. In this case, the surface tension is given by

$$\gamma = \frac{F}{2\ell}$$

We must use 2ℓ for the length because the surface film exerts forces on the inside and outside of the ring.

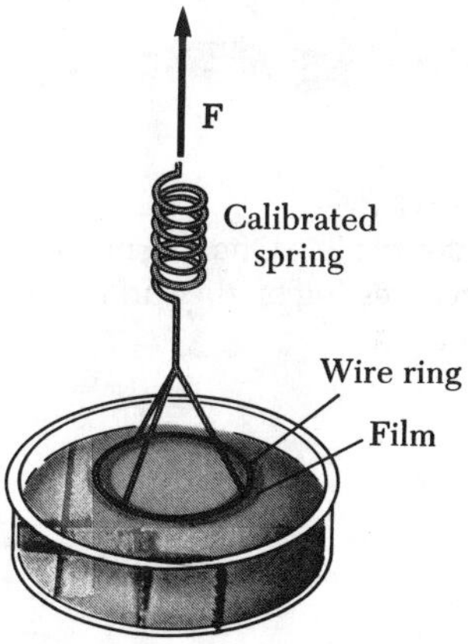

Figure 2.3
An apparatus for measuring the surface tension of liquids. The force on the wire ring is measured just before it breaks free of the liquid.

The surface tension of liquids decreases with increasing temperature. This occurs because the faster moving molecules of a hot liquid are not bound together as strongly as are those in a cooler liquid. Furthermore, certain ingredients added to liquids decrease surface tension. For example, soap or detergent decreases the surface tension of water. This reduction in surface tension makes it easier for soapy water to penetrate the cracks and crevices of your clothes to clean them better than plain water. An effect similar to this occurs in the lungs. The surface tissue of the air sacs in the lungs contains a fluid that has a surface tension of about 0.050 N/m. A liquid with a surface tension this high would make it very difficult for the lungs to expand as one inhales. However, as the area of the lungs increases with inhalation, the body secretes into the tissue a substance that gradually reduces the surface tension of the liquid. At full expansion, the surface tension of the lung fluid can drop to as low as 0.005 N/m.

Example 2.1 Walking on Water

In this example, we shall illustrate how an insect is supported on the surface of water by surface tension. Let us assume that the insect's "foot" is spherical. When the insect steps onto the water with all six legs, a depression is formed in the water around each foot, as shown in Figure 2.4. The surface tension of the water produces upward forces on the water which tend to restore the water surface to its normally flat shape. If the insect has a mass of 2×10^{-5} kg and if the radius of each foot is 1.5×10^{-4} m, find the angle θ.

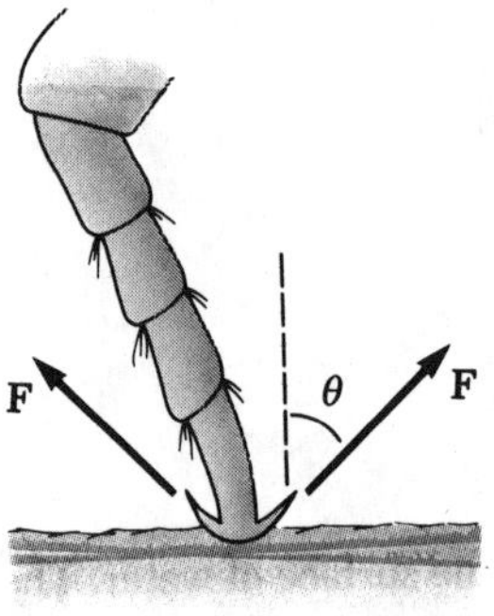

Figure 2.4
(Example 2.1) One foot of an insect resting on the surface of water.

Solution
From the definition of surface tension, we can find the net force F directed tangential to the depressed part of the water surface:

$$F = \gamma L$$

The length L along which this force acts is equal to the distance around the insect's foot, $2\pi r$. (It is assumed that the insect depresses the water surface such that the radius of the depression is equal to the radius of the foot.) Thus,

$$F = \gamma 2\pi r$$

and the net vertical force is

$$F_v = \gamma 2\pi r \cos\theta$$

Since the insect has six legs, this upward force must equal one sixth the weight of the insect, assuming its weight is equally distributed on all six feet. Thus,

$$\gamma 2\pi r \cos\theta = \frac{1}{6} w$$

(1)
$$\cos\theta = \frac{w}{12\pi r\gamma} = \frac{(2\times10^{-5}\ \text{kg})(9.80\ \text{m/s}^2)}{12\pi(1.5\times10^{-4}\ \text{m})(0.073\ \text{N/m})}$$

$$\theta = 62°$$

Note that if the weight of the insect were great enough to make the right side of (1) greater than unity, a solution for θ would be impossible because the cosine can never be greater than unity. Under these conditions, the insect would sink.

Capillarity

Capillary tubes are tubes in which the diameter of the opening is very small. In fact, the word capillary means "hair-like". If such a tube is inserted into a fluid for which adhesive forces dominate over cohesive forces, the liquid will rise into the tube, as shown in Figure 2.5. The rising of the liquid in the tube can be explained in terms of the shape of the surface of the liquid and in terms of the surface tension effects in the liquid. At the point of contact between liquid and solid, the upward force of surface tension is directed as shown in Figure 2.5. From Equation 2.1, the magnitude of this force is

$$F = \gamma L = \gamma(2\pi r)$$

We use $L = 2\pi r$ here because the liquid is in contact with the surface of the tube at all points around its circumference. The vertical component of this force due to surface tension is

$$F_v = \gamma(2\pi r)(\cos\phi) \qquad \textbf{(2.2)}$$

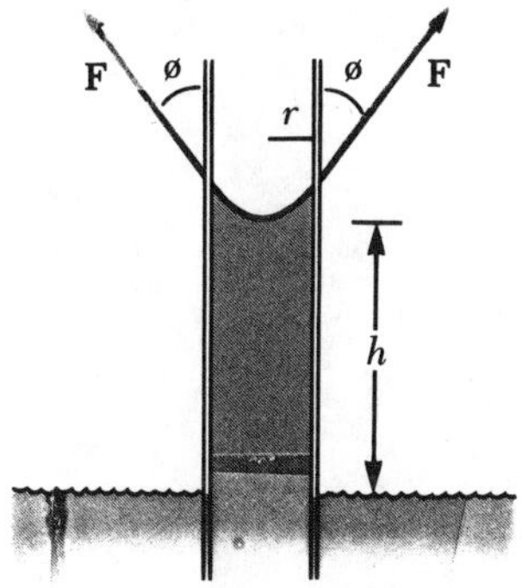

Figure 2.5
A liquid rises in a narrow tube because of capillary action, a result of surface tension and adhesive forces.

In order for the liquid in the capillary tube to be in equilibrium, this upward force must be equal to the weight of the cylinder of water of height h inside the capillary tube. The weight of this water is

$$w = Mg = \rho Vg = \rho g\pi r^2 \qquad \textbf{(2.3)}$$

Equating F_v in Equation 2.2 to w in Equation 2.3, we have

$$\gamma(2\pi r)(\cos\phi) = \rho g\pi r^2 h$$

Thus, the height to which water is drawn into the tube is

$$h = \frac{2\gamma}{\rho g r}\cos\phi \qquad \textbf{(2.4)}$$

If a capillary tube is inserted into a liquid in which cohesive forces dominate over adhesive forces, the level of the liquid in the capillary tube will be below the surface of the surrounding fluid, as shown in Figure 2.6. An analysis similar to that done above would show that the distance h the surface is depressed is given by Equation 2.4.

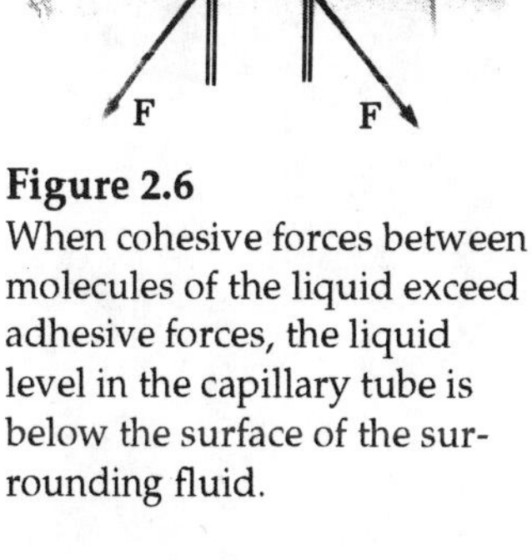

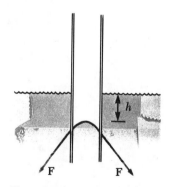

Figure 2.6
When cohesive forces between molecules of the liquid exceed adhesive forces, the liquid level in the capillary tube is below the surface of the surrounding fluid.

Capillary tubes are often used to draw small samples of blood from a needle prick in the skin. Capillary action must also be considered in the construction of concrete-block buildings because water seepage through capillary pores in the blocks or the mortar may cause damage to the inside of the building. To prevent this, the blocks are usually coated with a waterproofing agent either outside or inside the building. Water seepage through a wall is an undesirable effect of capillary action, but paper towels use capillary action in a useful manner to absorb spilled fluids.

Example 2.2 How High Does the Water Rise?

Find the height to which water would rise in a capillary tube with a radius equal to 5 x 10^{-5}m. Assume that the angle of contact between the water and the material of the tube is small enough to be considered zero.

Solution

The surface tension of water is 0.073 N/m. For a contact angle of 0°, we have cos ϕ = cos 0°= 1, so that Equation 2.4 gives

$$h = \frac{2\gamma}{\rho g r} = \frac{2(0.073\ \mathrm{N/m})}{(10^3\ \mathrm{kg/m^3})(9.80\ \mathrm{m/s^2})(5 \times 10^{-5}\ \mathrm{m})} = 0.29\ \mathrm{m}$$

PROBLEMS

SURFACE TENSION AND CAPILLARITY

1. A vertical force of 1.61×10^{-2} N is required to lift a wire ring of radius 1.75 cm from the surface of a container of blood plasma. Calculate the surface tension of blood plasma from this information.

2. Each of the six legs of an insect on a water surface makes a depression 0.25 cm in radius with a contact angle of 45°. Calculate the mass of the insect.

3. The surface tension of ethanol is 0.0227 N/m, and the surface tension of tissue fluid is 0.050 N/m. A force of 7.13×10^{-3} N is required to lift a 5 cm diameter wire ring vertically from the surface of ethanol. What should the diameter of the ring be so that this same force would lift it from the tissue fluid?

4. A certain fluid has a density of 1080 kg/m^3 and is observed to rise to a height of 2.1 cm in a 1 mm diameter tube. The contact angle between the wall and the fluid is zero. Calculate the surface tension of the fluid.

5. Whole blood has a surface tension of 0.058 N/m and a density of 1050 kg/m^3. To what height can whole blood rise in a capillary blood vessel that has a radius of 2×10^{-6} m if the contact angle is zero?

6. The density of water is 1000 kg/m^3 and its surface tension is 0.073 N/m at 20° C. Calculate the height to which water will rise in a capillary of diameter 10^{-4} m. Assume a contact angle of zero and a temperature of 20° C.

7. A staining solution used in a microbiology laboratory has a surface tension of 0.088 N/m and a density 1.035 times the density of water. What must the diameter of a capillary tube be so that this solution will rise to a height of 5 cm? (Assume a zero contact angle.)

8. A capillary tube 1 mm in radius is immersed in a beaker of mercury. The mercury level inside the tube is found to be 0.536 cm *below* the level of the reservoir. If the surface tension of mercury is 0.465 N/m, determine the contact angle between mercury and glass.

SOLUTIONS

SURFACE TENSION AND CAPILLARITY

1. Because there are two edges (the inside and outside of the ring) we have,

$\gamma = F/2L$

We have $L = 2\pi r = 2\pi(1.75 \times 10^{-2}\text{ m}) = 1.01 \times 10^{-1}\text{ m}$

and $\gamma = (1.61 \times 10^{-2}\text{ N})/2(1.01 \times 10^{-1}\text{ m}) = 7.32 \times 10^{-2}\text{ N/m}$

2. The force on one of the legs of the insect is $F = \gamma L$ where L is the circumference of the circular depression in which the foot rests;
$L = 2\pi r$. The vertical component of this force is
$F_v = F\cos 45° = \gamma L\cos 45°$

$= (7.3 \times 10^{-2}\text{ N/m})(2\pi)(2.5 \times 10^{-3}\text{ m})\cos 45° = 8.11 \times 10^{-4}\text{ N}$

The total support force will be $6F_v$, which is equal to the weight of the insect, w. Thus, we see that the mass of the insect is

$$m = \frac{6Fv}{g} = \frac{6(8.11 \times 10^{-4}\text{N})}{9.8\text{ m/s}^2} = 4.96 \times 10^{-4}\text{ kg} = 0.496\text{ g}$$

3. Because we have a two-sided surface, the force on the ring of radius r (diameter d) is

$F = 2\gamma L = 2\gamma(2\pi r) = 4\pi\gamma r = 2\pi\gamma d$

This force is the same for ethanol and the tissue fluid. Thus,

$2\pi\gamma_e d_e = 2\pi\gamma_f d_f$

or,

$$d_f = \frac{\gamma_e}{\gamma_f} d_e = \frac{0.0227}{0.050}(5\text{ cm}) = 2.27\text{ cm}$$

4. The vertical component of the surface force is equal to the weight of water inside the capillary tube.

$F_v = w$

$\gamma L\cos\theta = \gamma(2\pi r)\cos\theta = w = \rho(\pi r^2)hg$

Where h is the height of the water in the tube. We solve the above for the surface tension.

$$\gamma = \frac{\rho rgh}{2\cos\theta} = \frac{(1080\text{ kg/m}^3)(5 \times 10^{-4}\text{ m})(2.1 \times 10^{-2}\text{ m})(9.8\text{ m/s}^2)}{2} = 5.56 \times 10^{-2}\text{ N/m}$$

5. The height the blood can rise is given by

$$h = \frac{2\gamma\cos\theta}{\rho g r} = \frac{2(5.8 \times 10^{-2}\,\text{N/m})}{(1050\,\text{kg/m}^3)(9.8\,\text{m/s}^2)(2 \times 10^{-6}\,\text{m})} = 5.64\,\text{m}$$

6. We have

$$h = \frac{2\gamma\cos\theta}{\rho g r} = \frac{2(7.3 \times 10^{-2}\,\text{N/m})}{(1000\,\text{kg/m}^3)(9.8\,\text{m/s}^2)(5 \times 10^{-5}\,\text{m})} = 2.98 \times 10^{-1}\,\text{m} = 29.8$$

7. We have

$$h = 5 \times 10^{-2}\,\text{m} = \frac{2\gamma\cos\theta}{\rho g r} = \frac{2(8.8 \times 10^{-2}\,\text{N/m})}{(1035\,\text{kg/m}^3)(9.8\,\text{m/s}^2)r}$$

From which, we find $r = 3.47 \times 10^{-4}$ m, or a diameter of 0.694 mm.

8. Because the level of the fluid is below the level of the reservoir, h is a negative number. Thus, we find

$$h = 5.36 \times 10^{-3}\,\text{m} = \frac{2\gamma\cos\theta}{\rho g r} = \frac{2(0.465\,\text{N/m})\cos\theta}{(13.6 \times 10^3\,\text{kg/m}^3)(9.8\,\text{m/s}^2)(10^{-3}\,\text{m})}$$

This can be solved to find $\cos\theta = -0.768$
and $\theta = 140°$

2.2 Reynolds Number and Poiseuille's Law

Reynolds Number

At sufficiently high velocities, fluid flow changes from simple streamline flow to turbulent flow, that is, flow characterized by a highly irregular motion of the fluid. Experimentally it is found that the onset of turbulence in a tube is determined by a dimensionless factor called the **Reynolds number**, given by

$$RN = \frac{\rho v d}{\eta} \quad \textbf{(2.5)}$$

where ρ is the density of the fluid, v is the average velocity of the fluid along the direction of flow, d is the diameter of the tube, and η is the viscosity of the fluid. If RN is below about 2000, the flow of fluid through a tube is streamline; turbulence occurs if RN is above 3000. In the region between 2000 and 3000, the flow is unstable, meaning that the fluid can move in streamline flow but any small disturbances will cause its motion to change to turbulent flow.

Example 2.3 Turbulent Flow of Blood

Determine the velocity at which blood flowing through an artery of diameter 0.2 cm would become turbulent. Assume that the density of blood is 1.05×10^3 kg/m^3 and that its viscosity is 2.7×10^{-3} N• s/m^2.

Solution

At the onset of turbulence, the Reynolds number is 3000. Thus, the velocity of the blood would have to be

$$v = \frac{\eta(RN)}{\rho d} = \frac{(2.7 \times 10^{-3}\ \mathrm{N \cdot s/m^2})(3000)}{(1.05 \times 10^{3}\ \mathrm{kg/m^3})(0.2 \times 10^{-2}\ \mathrm{m})} = 3.86\ \mathrm{m/s}$$

Poiseuille's Law

Figure 2.7 shows a section of a tube containing a fluid under a pressure P_1 at the left end and a pressure P_2 at the right. Because of this pressure difference, the fluid will flow through the tube. The rate of flow (volume per unit time) depends on the pressure difference ($P_1 - P_2$), the dimensions of the tube, and the viscosity of the fluid. The relationship between these quantities was derived by a French scientist, J. L. Poiseuille (1799-1869), who assumed that the flow was streamline. His result, known as **Poiseuille's law**, is

$$\text{Rate of flow} = \frac{\Delta V}{\Delta t} = \frac{(P_1 - P_2)(\pi R^4)}{8L\eta} \quad \textbf{2.6}$$

where R is the radius of the tube, L is its length, and η is the coefficient of viscosity. We shall not attempt to derive this equation here because the methods of integral calculus are required. However, you should note that the equation does agree with common sense. That is, it is reasonable that the rate of flow should increase if the pressure difference across the tube or the tube radius increases. Likewise, the flow rate should decrease if the viscosity of the fluid or the length of the tube increases. Thus, the presence of R and the pressure difference in the numerator of Equation 2.6 and of L and η in the denominator makes sense.

From Poiseuille's law, we see that in order to maintain a constant flow rate, the pressure difference across the tube has to increase if the viscosity of the fluid increases. This is important when one considers the flow of blood through the circulatory system. The viscosity of blood increases as the number of red blood cells rises. Thus, blood with a high concentration of red blood cells requires greater pumping pressure from the heart to keep it circulating than does blood of lower red blood cell concentration.

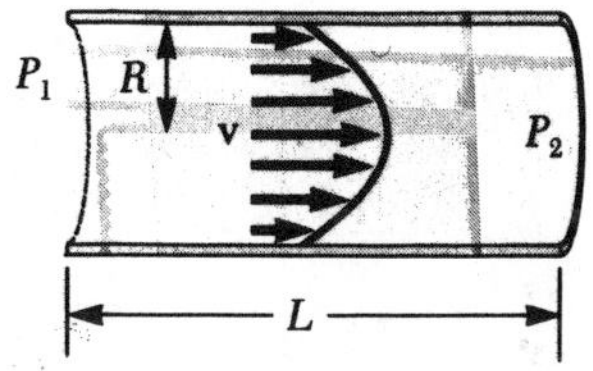

Figure 2.7
Velocity profile of a fluid flowing through a uniform pipe of circular cross-section. The rate of flow is given by Poiseuille's law. Note that the fluid velocity is greatest at the middle of the pipe.

Note that the flow rate varies as the radius of the tube raised to the fourth power. Consequently, if a constriction occurs in a vein or artery, the heart will have to work considerably harder in order to produce a higher pressure drop and hence to maintain the required flow rate.

Example 2.4 A Blood Transfusion

A patient receives a blood transfusion through a needle of radius 0.2 mm and length 2 cm. The density of blood is 1050 kg/m^3. The bottle supplying the blood is 0.5 m above the patient's arm. What is the rate of flow through the needle?

Solution
The pressure differential between the level of the blood and the patient's arm is

$$P_1 - P_2 = \rho g h = (1050\ \mathrm{kg/m^3})(9.80\ \mathrm{m/s^2})(0.5\ \mathrm{m}) = 5.15 \times 10^3\ \mathrm{Pa}$$

Thus, the rate of flow, from Poiseuille's law, is

$$\frac{\Delta V}{\Delta t} = \frac{(P_1 - P_2)(\pi R4)}{8L\eta} = \frac{(5.15 \times 10^3\ \mathrm{Pa})(\pi)(2 \times 10^{-4}\ \mathrm{m})^4}{8(2 \times 10^{-2}\ \mathrm{m})(2.7 \times 10^{-3}\ \mathrm{N \cdot s/m^2})} = 5.98 \times 10^{-8}\ \mathrm{m^3/s}$$

Exercise How long will it take to inject 1 pint (500 cm^3) of blood into the patient?

Answer 139 min.

PROBLEMS

REYNOLDS NUMBER AND POISEUILLE'S LAW

1. A block of ice (temperature 0°C) is drawn over a level surface lubricated by a layer of water 0.1 mm thick. The face of the block in contact with the surface has dimensions 1.20 m by 0.80 m. Determine the magnitude of the force needed to pull the block with a constant speed of 0.5 m/s. At 0°C, the viscosity of water is $\eta = 1.79 \times 10^{-3}$ N s/m^2.

2. A metal block is pulled over a horizontal surface that has been coated with a layer of lubricant 1.00 mm thick. The face of the block in contact with the surface has dimensions 0.40 m by 0.12 m. A force of 1.9 N is required to move the block at a constant speed of 0.5 m/s. Calculate the coefficient of viscosity of the lubricant.

3. A thin 1.5 mm coating of glycerin has been placed between two microscope slides of width 1 cm and length 4 cm. Find the force required to pull one of the microscope slides at a speed of 0.3 m/s relative to the other.

4. A straight horizontal pipe with a diameter of 1 cm and a length of 50 m carries oil with a coefficient of viscosity of 0.12 Pa · s. At the output of the pipe, the flow rate is 8.6×10^{-5} m^3/s and the pressure is 1 atm. Find the gauge pressure at the pipe input.

5. A hypodermic needle is 3 cm in length and 0.3 mm in diameter. What excess pressure is required along the needle so that the flow rate of water through it will be 1 g/s? (Use 10^{-3} Pa · s as the viscosity of water.)

6. A needle of radius 0.3 mm and length 3 cm is used to give a patient a blood transfusion. Assume the pressure differential across the needle is achieved by elevating the blood 1 m above the patient's arm (a) What is the rate of flow of blood through the needle? (b) At this rate of flow of blood, how long will it take to inject 1 pint (approximately 500 cm^3) of blood into the patient?
Density of blood = 1050 kg/m^3, and its coefficient of viscosity = 4×10^{-3} N•s/m^2.

7. The water pressure in a horizontal pipe decreases 0.5 atm per 100 m when the flow rate in the pipe is 30.0 liter/min. Determine the radius of the pipe. (Assume a temperature of 20°C.)

8. What diameter needle should be used to inject a volume of 500 cm^3 of a solution into a patient in 30 min? Assume that the needle length is 2.5 cm and that the solution is elevated 1 m. Furthermore, assume the viscosity and density of the solution as those of pure water.

9. The pulmonary artery, which connects the heart to the lungs, has an inner radius of 2.6 mm and is 8.4 cm long. If the pressure drop between the heart and lungs is 400 Pa, what is the average speed of blood in the pulmonary artery? (See problem 6 above for the physical properties of blood.)

10. A pipe carrying 20°C water has a diameter of 2.5 cm. Estimate the maximum flow speed if the flow is to be laminar.

11. What is the Reynolds number for the flow of liquid in the 1.2 m diameter Alaska pipeline? The density of crude oil is 850 kg/m^3, its velocity is 3 m/s, and its speed is 0.3 Pa·s. Is the flow laminar or turbulent?

12. Determine the speed at which the flow of water through a 0.5 cm diameter pipe will become turbulent. (RN > 3000).

13. Assume a value of 980 for the Reynolds Number for blood in an artery and a viscosity of 4×10^{-3} N s/m^2 for whole blood. If the density of whole blood is equal to 1.05×10^3 kg/m^3, at what velocity does blood flow through an artery 0.45 cm in diameter?

14. The aorta in humans has a diameter of about 2 cm and, at certain times, the blood speed through it is about 55 cm/s. Is the blood flow turbulent? (Values for several blood properties that you may need are given in problem 6 above.)

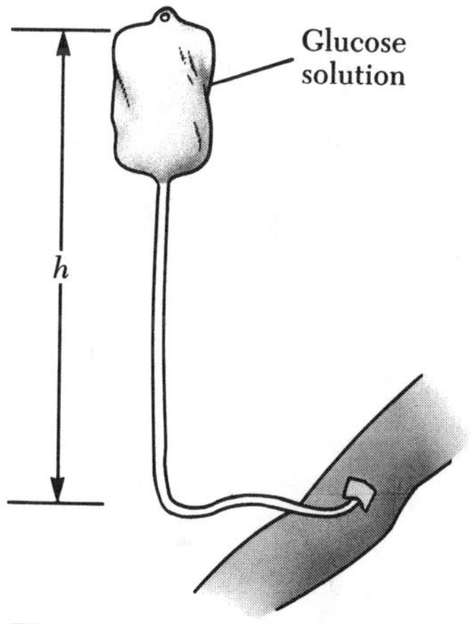

Figure 2.8

15. A collapsible plastic bag, (Fig. 2.8) contains a glucose solution. If the average gauge pressure in the artery is 1.33×10^4 Pa, what must be the minimum height h of the bag in order to infuse glucose into the artery? Assume the specific gravity of the solution is 1.02.

16. Blood of density 1050 kg/m^3 is to be administered to a patient. To do so the blood is raised to a height of about 1 m higher than the level of the patient's arm. How much greater is the pressure of the blood at this level than if the container were at the same level as the arm?

SOLUTIONS

REYNOLDS NUMBER AND POISEUILLE'S LAW

1. From the definition of the coefficient of viscosity, we have

$$F = \frac{\eta Av}{L} = \frac{(1.79 \times 10^{-3}\ \text{N} \cdot \text{s/m}^2)(0.8\ \text{m})(1.2\ \text{m})(0.5\ \text{m/s})}{10^{-4}\ \text{m}} = 8.59\ \text{N}$$

2. $$\eta = \frac{FL}{Av} = \frac{(1.9\ \text{N})(10^{-3}\ \text{m})}{(4.8 \times 10^{-2}\ \text{m})(0.5\ \text{m/s})} = 7.92\ \text{N} \times 10^{-2}\ \text{N} \cdot \text{s/m}^2$$

3. From the definition of the coefficient of viscosity, we have

$$F = \frac{\eta Av}{L} = \frac{(1500 \times 10^{-3}\ \text{N} \cdot \text{s/m}^2)(4 \times 10^{-4}\ \text{m}^2)(0.3\ \text{m/s})}{1.5 \times 10^{-3}\ \text{m}} = 0.120\ \text{N}$$

4. From Poiseuille's law

$$P_1 - P_2 = \frac{(\text{flow rate})8\eta L}{\pi R^4}$$

$$= \frac{(8.6 \times 10^{-5}\ \text{m}^3/\text{s})8(0.12\ \text{N} \cdot \text{s/m}^2)(50\ \text{m})}{\pi(5 \times 10^{-3}\text{m})^4} = 2.1 \times 10^6\ \text{Pa} = 20.7\ \text{atm.}$$

Also, since $P_2 = 1$ atm, this is also the gauge pressure at the inlet point of the pipe.

5. Poiseuille's law gives the flow rate as

$$\text{Flow rate} = \frac{(\Delta P)\pi R^4}{8\eta L}$$

$$\text{Therefore } \Delta P = \frac{(\text{flow rate})8\eta L}{\pi R^4}$$

$$= \frac{(10^{-6}\ \text{m}^3/\text{s})8(0.03\ \text{m})(1 \times 10^{-3}\ \text{N} \cdot \text{s/m}^2)}{\pi(1.5 \times 10^{-4}\ \text{m})^4}$$

$$= 1.5 \times 10^5\ \text{Pa}$$

6. **(a)** The pressure differential across the needle is

$$\Delta P = \rho g y = (1050\ \text{kg/m}^3)(9.8\ \text{m/s}^2)(1.0\ \text{m}) = 1.03 \times 10^4\ \text{Pa}.$$

Then we find the flow rate as

$$\text{Flow rate} = \frac{(\Delta P)\pi R^4}{8\eta L} = \frac{(1.03 \times 10^4\ \text{Pa})\pi(3 \times 10^{-4}\ \text{m})^4}{8(3 \times 10^{-2}\ \text{m})(4 \times 10^{-3}\ \text{N}\cdot\text{s/m}^2)}$$

$$= 2.73 \times 10^{-7}\ \text{m}^3/\text{s} = 2.73 \times 10^{-1}\ \text{cm}^3/\text{s}$$

(b) At this flow rate, the time to inject 500 cm³ is

$$\Delta t = \frac{500\ \text{cm}^3}{0.273\ \text{cm}^3/\text{s}} = 1.83 \times 10^3\ \text{s} = 30.6\ \text{min}.$$

7. flow rate = 30.0 liter/min = $5 \times 10^{-4}\ \text{m}^3/\text{s}$

From Poiseuille's law, we have

$$R^4 = \frac{(\text{rate})8L\eta}{\pi(\Delta P)} = \frac{(5 \times 10^{-4}\ \text{m}^3/\text{s})8(10^{-3}\ \text{Pa}\cdot\text{s})(10^2\ \text{m})}{\pi(5.06 \times 10^4\ \text{Pa})}$$

Which gives $R = 7.08$ mm

8. The required flow rate is

$$\text{flow rate} = \frac{500\ \text{cm}^3}{1800\ \text{s}} = 2.78 \times 10^{-1}\ \text{cm}^3/\text{s} = 2.78 \times 10^{-7}\ \text{m}^3/\text{s}$$

If the solution is elevated 1 m, the pressure differential across the needle is

$$\Delta P = \rho g y = (1000\ \text{kg/m}^3)(9.8\ \text{m/s}^2)(1.0\ \text{m}) = 9800\ \text{Pa}$$

We find the radius using Poiseuille's law:

$$R^4 = \frac{(\text{rate})8L\eta}{\pi(\Delta P)}$$

$$= \frac{8(2.5 \times 10^{-2}\ \text{m})(1 \times 10^{-3}\ \text{N}\cdot\text{s/m}^2)(2.78\ 10^{-7}\ \text{m}^3/\text{s})}{\pi(9800\ \text{Pa})}$$

From which we find

$R = 2.06 \times 10^{-4}$ m = 0.206 mm, or diameter = 0.412 mm.

9.

$$\text{Flow rate} = \frac{(\Delta P)\pi R^4}{8L\eta} = \frac{(400\ \text{Pa})\pi(2.6 \times 10^{-3}\ \text{m})^4}{8(4 \times 10^{-3}\ \text{Pa}\cdot\text{s})(8.4 \times 10^{-2}\ \text{m}) = 2.14 \times 10^{-5}\ \text{m}^3/\text{s}} = 2.14 \times 10^{-7}\ \text{m}^3/\text{s}$$

$$\text{Then } v = \frac{\text{flow rate}}{\text{area}} = \frac{2.14 \times 10^{-5}\ \text{m}^3/\text{s}}{\pi(2.6 \times 10^{-3}\ \text{m})^2}$$

10. From the definition of the Reynolds number,

$$\text{Then } v_{\text{max}} = \frac{(\text{RN})\eta}{\rho d} = \frac{(2000)(10^{-3}\ \text{N}\cdot\text{s/m}^2)}{(10^3\ \text{kg/m}^3)(2.5 \times 10^{-2}\ \text{m})} = 8 \times 10^{-2}\ \text{m/s} = 8\ \text{cm/s}$$

11. From the definition of the Reynolds number,

$$\text{RN} = \frac{\rho v d}{\eta} = \frac{(850\ \text{kg/m}^3)(3\ \text{m/s})(1.2\ \text{m})}{(0.3\ \text{N}\cdot\text{s/m}^2)} = 10{,}200$$

The onset of turbulence is at RN about 3000, so this is definitely turbulent.

12. The minimum value of the Reynolds number for turbulent flow is 3000. From the definition of the Reynolds number, we find the minimum velocity for turbulent flow as

$$v = \frac{\eta(\text{RN})}{\rho d} = \frac{(10^{-3}\ \text{N}\cdot\text{s/m}^2)(3000)}{(10^3\ \text{kg/m}^3)(5 \times 10^{-3}\ \text{m})} = 0.6\ \text{m/s}$$

13. The velocity of the blood is found from the definition of the Reynolds number as,

$$v = \frac{\eta(\text{RN})}{\rho d} = \frac{(4 \times 10^{-3}\ \text{N}\cdot\text{s/m}^2)(980)}{(1050\ \text{kg/m}^3)(4.5 \times 10^{-3}\ \text{m})} = 0.830\ \text{m/s}$$

14. The Reynolds number is

$$\text{RN} = \frac{\rho v d}{\eta} = \frac{(1050\ \text{kg/m}^3)(0.55\ \text{m/s})(2 \times 10^{-2}\ \text{m})}{(4 \times 10^{-3}\ \text{N}\cdot\text{s/m}^2)} = 2890$$

In this region the flow is unstable, but not necessarily turbulent

15. The gauge pressure of the fluid at the level of the needle must equal the gauge pressure in the artery.

$$\rho gh = 1.333 \times 10^{4} \text{ Pa}$$

$$h = \frac{1.333 \times 10^{4} \text{ Pa}}{(1.02 \times 10^{3} \text{ kg/m}^3)(9.8 \text{ m/s}^2)} = 1.33 \text{ m}$$

16. We use $\Delta P = \rho gh$, where h is the height above the arm. Thus, $\Delta P = (1050 \text{ kg/m}^3)(9.8 \text{ m/s}^2)(1 \text{ m}) = 1.03 \times 10^4$ Pa.

2.3 Transport Phenomena

When a fluid flows through a tube, the basic mechanism that produces the flow is a difference in pressure across the ends of the tube. This pressure difference is responsible for the transport of a mass of fluid from one location to another. The fluid may also move from place to place because of a second mechanism, one that depends on a concentration difference between two points in the fluid, as opposed to a pressure difference. When the concentration (the number of molecules per unit volume) is higher at one location than at another, molecules will flow from the point where the concentration is high to the point where it is lower. The two fundamental processes involved in fluid transport resulting from concentration differences are called *diffusion* and *osmosis*.

Diffusion

You can imagine what happens when someone wearing a strong shaving lotion or perfume strolls into a crowded room. All eyes turn to seek out the source of the delightful smell. The aroma spreads through the room by a process called diffusion.

> **In a diffusion process, molecules move from a region where their concentration is high to a region where their concentration is lower.**

That is, the molecules of the lotion or perfume move from the source (near the person's face), where there are many molecules per unit volume, throughout the room, to regions where the concentration of these molecules is lower. Although the example used here is one of diffusion in air, the process also occurs in liquids and, to a lesser extent, in solids. For example, if a drop of food coloring is placed in a glass of water, the coloring soon spreads throughout the liquid by diffusion. In either case, diffusion ceases when there is a uniform concentration at all locations in the fluid.

To understand why diffusion occurs, consider Figure 2.9, which represents a the left side. For example, this could be accomplished by releasing a few drops of perfume into the left side of the container. The dashed line in Figure 2.9 represents an imaginary barrier separating the region of high concentration from the region of lower concentration. Because the molecules are moving with high speeds in random directions, many of them will cross the imaginary barrier moving from left to right. Very few molecules of perfume will pass through this area moving from right to left simply because there are very few of them on the right side of the container at any instant. Thus, there will always be a *net* movement from the region where there are many molecules to the region where there are fewer molecules. For this reason, the concentration on the left side of the container will decrease in time and that on the right side will increase. There will be no *net* movement across the cross-sectional area once the concentration is the same on both sides, the number of molecules diffusing from right to left in a given time interval will equal the number moving from left to right in the same time interval.

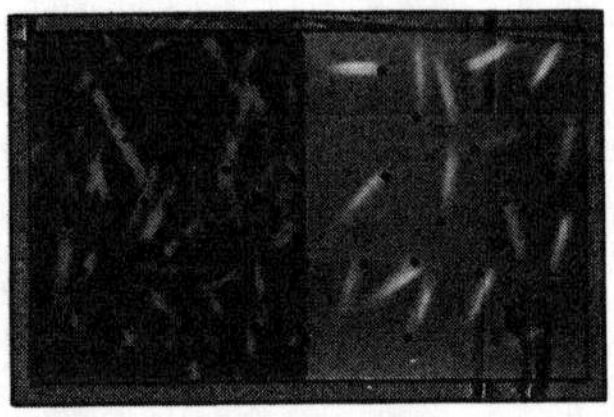

Figure 2.9
When concentration of gas molecules on the left side of the container exceeds the concentration on the right side, there will be a net motion (diffusion) of molecules from left to right.

The basic equation for diffusion is Fick's law, which in equation form is

$$\text{Diffusion rate} = \frac{\text{mass}}{\text{time}} = \frac{\Delta M}{\Delta t} = DA\left(\frac{C_2 - C_1}{L}\right) \qquad \textbf{(2.7)}$$

where D is a constant of proportionality. The left side of this equation is called the diffusion rate and is a measure of the mass being transported per unit time. This equation says that

the rate of diffusion is proportional to the cross-sectional area. A and to the change in concentration per unit distance, $(C_2 - C_1)/L$, which is called the concentration gradient.

The concentrations C_1 and C_2 are measured in kilograms per cubic meter. The proportionality constant D is called the **diffusion coefficient** and has units of square meters per second. Table 2.2 lists diffusion coefficients for a few substances.

Table 2.2
Diffusion Coefficients for Various Substances at 20° C

Substance	$D(m^2/s)$
Oxygen through air	6.4×10^{-5}
Oxygen through tissue	1×10^{-11}
Oxygen through water	1×10^{-9}
Sucrose through water	5×10^{-10}
Hemoglobin through water	76×10^{-11}

The Size of Cells and Osmosis

Diffusion through cell membranes is extremely vital in carrying oxygen to the cells of the body and in removing carbon dioxide and other waste products from them. Oxygen is required by the cells for those metabolic processes in which substances are either synthesized or broken down. In such metabolic processes, the cell uses up oxygen and produces carbon dioxide as a by-product. A fresh supply of oxygen diffuses from the blood, where its concentration is high, into the cell, where its concentration is low. Likewise, carbon dioxide diffuses from the cell into the blood, where it is in lower concentration. Water, ions, and other nutrients also pass into and out of cells by diffusion.

A common characteristic of cells in all plants and animals is their extremely small size. The adult human body contains literally trillions of cells. In order to understand why cells are so small, we must consider the relationship between the surface area of an object and its volume.

Let us consider a cube 2 cm on a side. The area of one of its faces is 2 cm $\times$ 2 cm= 4 cm^2, and because a cube has six sides, the total surface area is 24 cm^2. Its volume is 2 cm $\times$ 2 cm $\times$ 2 cm = 8 cm^3. Hence, the ratio of surface area to volume is 24 / 8 = 3. Now consider a larger cube, one measuring 3 cm on a side. Repeating the calculations gives us a surface area of 54 cm^2 and a volume of 27 cm^3. In this case, the ratio of surface area to volume is 54 / 27 = 2. Thus, we see that as the size of an object decreases, the ratio of its surface area to its volume increases. This, of course, says that a small cell has a larger surface-area-to-volume ratio than a large cell. But how does this pertain to the operation of a cell?

A cell can function properly only if it can (a) rapidly receive vital substances such as oxygen and (b) rapidly eliminate waste products. If such substances are to readily move into and out of cells, the cells should have a large surface area. However, if the volume of the cell is too large, it could take a considerable period of time for the nutrients to diffuse into the interior of the cell where they are needed. Under optimum conditions, the surface area of the cell should be large enough so that the exposed membrane area can exchange materials effectively while at the same time the volume should be small enough so that materials can reach or leave particular locations rapidly. To reach these optimum conditions, a small cell with its high surface-area-to-volume ratio is necessary.

Osmosis

As we have seen, the movement of material through cell membranes is necessary for the efficient functioning of cells. The diffusion of material through a membrane is partially determined by the size of the pores (holes) in the membrane wall. That is, small molecules, such as water, may pass through the pores easily while larger molecules, such as sugar, may pass through only with difficulty or not at all. A membrane that allows passage of some molecules but not others is called a selectively permeable membrane.

> **Osmosis is defined as the movement of water from a region where its concentration is high, across a selectively permeable membrane, into a region where its concentration is lower.**

As in the case of diffusion, osmosis continues until the concentrations on the two sides of the membrane are equal. Osmosis is often described simply as the diffusion of water across a membrane.

To understand the effect of osmosis on living cells, let us consider a particular cell in the body that contains a sugar concentration of 1%. (That is, 1 g of sugar is dissolved in enough water to make 100 ml of solution.) Now assume that this cell is immersed in a 5% sugar solution (5 g of sugar dissolved in enough water to make 100 ml). In such a situation, water would diffuse from inside the cell, where its concentration is higher, across the cell wall membrane, to the outside solution, where the concentration of water is lower. This loss of water from the cell would cause it to shrink and perhaps become damaged through dehydration. If the concentrations were reversed, water would diffuse into the cell, causing it to swell and perhaps burst. It should be obvious from this description that normal osmotic relationships must be maintained in the body. If solutions are introduced into the body intravenously, care must be taken to ensure that these solutions do not disturb the osmotic balance of the body because such a disturbance could lead to cell damage. For example, if 9% saline solution surrounds a red blood cell, the cell will shrink. On the other hand, if the saline solution is about 1%, the cell will eventually burst.

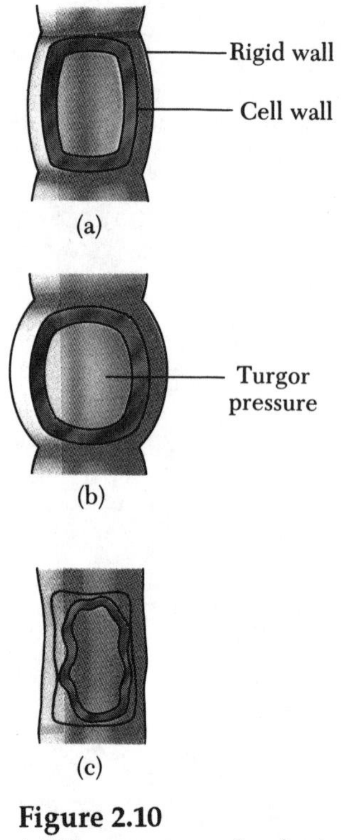

Figure 2.10
(a) The structure of a plant cell. (b) As water accumulates in its interior, the cell expands under turgor pressure. (c) When water in the cell's interior is depleted, the walls of the cell collapse.

Under normal circumstances, the cells of our bodies are in an environment such that there is no net movement of water into or out of them. However, certain one-celled organisms, such as protozoa, do not enjoy this osmotic equilibrium. These organisms usually live in fresh water, which obviously has a higher concentration of water than the solution inside the cell. To prevent an inflow of water to the point of bursting, these organisms possess an organelle (a tiny organ) that acts as a pump and continually forces water out of the cell.

Most plant cells are contained within a rigid wall, as shown in Figure 2.10. If water accumulates in the cell, it expands (Fig. 2.10(b)) and exerts a pressure, called turgor pressure, against the rigid wall. The rigidity of the wall prevents the cell from bursting. If water within the cell is depleted, the rigid wall collapses inward slightly, as in Figure 2.10(c). This causes the plant to wilt.

PROBLEMS

TRANSPORT PHENOMENA

1. Sucrose is allowed to diffuse along a 10 cm length of tubing filled with water. The tube is 6 cm^2 in cross-sectional area. The diffusion coefficient for this case is 5×10^{-10} m^2/s, and 8×10^{-14} kg is transported along the tube in a time of 15 s. What is the difference in the concentration levels of sucrose at the two ends of the tube?

2. In a diffusion experiment, it is found that, in 60 s, 3×10^{-13} kg of sucrose will diffuse along a horizontal pipe of cross-sectional area 1 cm^2. If the diffusion coefficient is 5×10^{-10} m^2/s, what is the concentration gradient (that is, the change in concentration per unit length along the path)?

3. Glycerine in water diffuses along a horizontal column that has a cross-sectional area of 2 cm^2. The concentration gradient is 3×10^{-2} kg/m^4, and the diffusion rate is found to be 5.7×10^{-15} kg/s. Determine the diffusion coefficient.

4. Use the data for sucrose in problem 1 to calculate how much sucrose will diffuse down a horizontal pipe of cross-sectional area 4 cm^2 in 10 s if the concentration gradient is 0.2 kg/m^4.

SOLUTIONS

TRANSPORT PHENOMENA

1. Fick's law enables us to find the difference in concentration as

$$\Delta C = \frac{(\text{diffusion rate})L}{DA} = \frac{(5.33 \times 10^{-15}\ \text{kg/s})(0.1\ \text{m})}{(5 \times 10^{-10}\ \text{m}^2/\text{s})(6 \times 10^{-4}\ \text{m}^2)} = 1.78 \times 10^{-3}\ \text{m}^3$$

2. From Fick's law, we find the concentration gradient by

$$\frac{\Delta C}{L} = \frac{(\text{diffusion rate})}{DA} = \frac{(5 \times 10^{-15}\ \text{kg/s})}{(5 \times 10^{-10}\ \text{m}^2/\text{s})(10^{-4}\ \text{m}^2)} = 0.10\ \text{kg/m}^4.$$

3. We use Fick's law to find the diffusion coefficient.

$$D = \frac{(\text{diffusion rate})}{A\left(\frac{\Delta C}{L}\right)} = \frac{(5.7 \times 10^{-15}\ \text{kg/s})}{(2 \times 10^{-4}\ \text{m}^2)(3 \times 10^{-2}\ \text{kg/m}^4)} = 9.50 \times 10^{-10}\ \text{m}^2/\text{s}$$

4. We have

$$\text{Diffusion rate} = DA\,\frac{\Delta C}{L} = (5 \times 10^{-10}\ \text{m}^2/\text{s})(4 \times 10^{-4}\ \text{m}^2)(0.2\ \text{kg/m}^4)$$
$$= 4 \times 10^{-14}\ \text{kg/s}$$

In 10 s the amount diffused is

$$\Delta m = (\text{diffusion rate})\Delta t = 4.00 \times 10^{-13}\ \text{kg}$$

2.4 Motion Through a Viscous Medium

When an object falls through air, its motion is impeded by the force of air resistance. In general, this force is dependent on the shape of the falling object and on its velocity. This viscous drag acts on all falling objects, but the exact details of the motion can be calculated only for a few cases in which the object has a simple shape, such as a sphere. In this section, we shall examine the motion of a tiny spherical object falling slowly through a viscous medium.

In 1845 a scientist named George Stokes found that the magnitude of the resistive force on a very small spherical object of radius r falling slowly through a fluid viscosity η is given by

$$F_r = 6\pi\eta r v \qquad \textbf{(2.8)}$$

This equation, called **Stoke's law**, has many important applications. For example, it describes the sedimentation of particulate matter in blood samples. It was used by Robert Millikan (1886–1953) to calculate the radius of charged oil droplets falling through air. From this, Millikan was ultimately able to determine the smallest known unit of electric charge. Millikan was awarded the Nobel prize in 1923 for this pioneering work on elemental charge.

As a sphere falls through a viscous medium, three forces act on it, as shown in Figure 2.11: $\mathbf{F}_r$ is the force of frictional resistance, **B** is the buoyant force of the fluid, and **w** is the weight of the sphere, whose magnitude is given by

$$w = \rho g V = \rho g\left(\frac{4}{3}\pi r^3\right)$$

where ρ is the density of the sphere and $\frac{4}{3}\pi r^3$ is its volume. According to Archimedes' principle, the magnitude of the buoyant force is equal to the weight of the fluid displaced by the sphere:

$$B = \rho_f g V = \rho_f g\left(\frac{4}{3}\pi r^3\right)$$

where ρ_f is the density of the fluid.

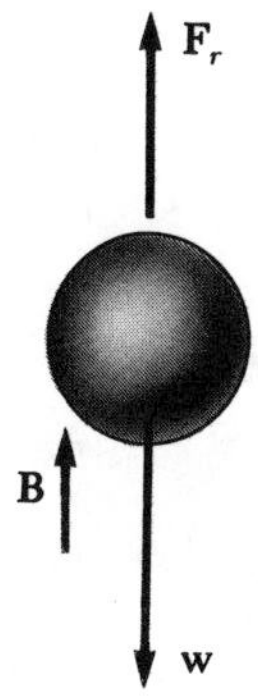

Figure 2.11
A sphere falling through a viscous medium. The forces acting on the sphere are the resistive frictional force $\mathbf{F}_r$, the buoyant force **B**, and the weight of the sphere **w**.

At the instant the sphere begins to fall, the force of frictional resistance is zero because the speed of the sphere is zero. As it accelerates, the speed increases and so does F_r. Finally, at a speed called the **terminal speed** v_t, *the resultant force goes to zero.* This occurs when the net upward force balances the downward weight force. Hence, the sphere reaches terminal speed when

$$F_r + B = w$$

or

$$6\pi\eta r v_t + \rho_f g\left(\frac{4}{3}\pi r^3\right) = \rho g\left(\frac{4}{3}\pi r^3\right)$$

When this is solved for v_t, we get

$$v_t = \frac{2r^2 g}{9\eta}(\rho - \rho_f) \qquad \textbf{(2.9)}$$

Example 2.5 A Falling Pearl

A pearl of density 2×10^3 kg /m^3 and radius 2 mm falls through a liquid shampoo of density 1.4×10^3 kg /m^3 and viscosity 500×10^{-3} N · s/m^2. Find the terminal speed of the pearl.

Solution

Substituting the values given into Equation 2.9, we have

$$v_t = \frac{2r^2 g}{9\eta}(\rho - \rho_f)$$

$$= \frac{2(2 \times 10^{-3}\text{ m})^2(9.80\text{ m/s}^2)}{9(500 \times 10^{-3}\text{ N} \cdot \text{s/m}^2)}(2 \times 10^3\text{ kg/m}^3 - 1.4 \times 10^3\text{ kg/m}^3)$$

$$= 1.05 \times 10^{-2}\text{ m/s}$$

PROBLEMS

MOTION THROUGH A VISCOUS MEDIUM

1. Small spheres of diameter 1 mm fall through water with a terminal speed of 1.1 cm/s. Calculate the density of the spheres.

2. A test tube 10 cm tall is filled with water that contains spherical particles of density 1.8 g/cm^3 in suspension. If the radius of the particle is 2×10^{-4} cm, find the time it takes a particle near the top of the tube to reach the bottom.

3. Calculate the viscous force on a spherical oil droplet that is 2×10^{-5} cm in diameter and falling with a speed of 0.4 mm/s in air. (Use $\eta = 1.8 \times 10^{-5}$ N s/m^2 as the coefficient of viscosity of air.)

4. The viscous force on an oil drop is measured to be 3×10^{-13} N when the drop is falling through air with a speed of 4.5×10^{-4} m/s. If the radius of the drop is equal to 2.5×10^{-6} m, what is the viscosity of air?

5. An oil drop ($\rho = 800$ kg/m^3) is falling with a terminal speed of 0.04 mm/s in air of density 1.29 kg/m^3. Calculate the radius of the oil drop. (The coefficient of viscosity of air is $\eta = 1.8 \times 10^{-5}$ N•s/m^2.)

6. Spherical particles of a protein of density 1.8 g/cm^3 are shaken up in a solution of water. The solution is allowed to stand for 1 h. If the depth of water in the tube is 5 cm, find the radius of the largest particles still in solution at the end of the hour.

SOLUTIONS

MOTION THROUGH A VISCOUS MEDIUM

1. We use

$$v_t = \frac{2r^2g}{9\eta}(\rho - \rho_f)$$

or

$$(\rho - \rho_f) = \frac{9\eta\, v_t}{2r^2g} = \frac{9(1 \times 10^{-3}\ \mathrm{N \cdot s/m^2})(1.1 \times 10^{-2}\ \mathrm{m/s})}{2(5 \times 10^{-4})^2(9.8\ \mathrm{m/s^2})} = 20.2\ \mathrm{kg/m^3}$$

Therefore, $\rho = \rho_f + 20.2\ \mathrm{kg/m^3} = 1000\ \mathrm{kg/m^3} + 20.2\ \mathrm{kg/m^3} = 1.02 \times 10^3\ \mathrm{kg/m^3}$

2. The terminal speed is

$$v_t = \frac{2r^2g}{9\eta}(\rho - \rho_f) = \frac{2(2 \times 10^{-6}\ \mathrm{m})^2(9.8\ \mathrm{m/s^2})}{9(1 \times 10^{-3}\ \mathrm{N \cdot s/m^2})}(1800\ \mathrm{kg/m^3} - 1000\ \mathrm{kg/m^3})$$

$$= 6.97 \times 10^{-6}\ \mathrm{m/s}$$

Thus, the time is

$$t = d/v_t = (1 \times 10^{-1}\ \mathrm{m})/(6.97 \times 10^{-6}\ \mathrm{m/s}) = 1.43 \times 10^4\ \mathrm{s} = 3.99\ \mathrm{h}$$

3. $F = 6\pi\eta r v = 6\pi(1.8 \times 10^{-5}\ \mathrm{N \cdot s/m^2})(1 \times 10^{-7}\ \mathrm{m})(4 \times 10^{-4}\ \mathrm{m/s}) = 1.36 \times 10^{-14}\ \mathrm{N}$

4. From $F = 6\pi\eta r v$, we have

$$\eta = \frac{F}{6\pi r v} = \frac{3 \times 10^{-13}\ \mathrm{N}}{6\pi(2.5 \times 10^{-6}\ \mathrm{m})(4.5 \times 10^{-4}\ \mathrm{m/s})} = 1.41 \times 10^{-5}\ \mathrm{N \cdot s/m^2}$$

5. From $v_t = \frac{2r^2g}{9\eta}(\rho - \rho_f)$ we find

$$r^2 = \frac{9\eta v_t}{2g(\rho - \rho_f)} = \frac{9(1.8 \times 10^{-5}\ \mathrm{N \cdot s/m^2})(4 \times 10^{-5}\ \mathrm{m/s})}{2(9.8\ \mathrm{m/s^2})(800 - 1.29)\ \mathrm{kg/m^3}}$$

gives $r = 6.43 \times 10^{-7}\ \mathrm{m} = 0.643\ \mu\mathrm{m}$

6. If at the end of one hour a particle is still in suspension, then its terminal speed must be less than 5 cm/h = 1.39 × 10^{-5} m/s. Thus, we use

$$v_t = \frac{2r^2 g}{9\eta}(\rho - \rho_f) \quad \text{to find}$$

$$r^2 = \frac{9\eta v_t}{2g(\rho - \rho_f)} = \frac{9(10^{-3}\ \mathrm{N \cdot s/m^2})(1.39 \times 10^{-5}\ \mathrm{m/s})}{2(9.8\ \mathrm{m/s^2})(800\ \mathrm{kg/m^3})}$$

and $r = 2.82 \times 10^{-6}$ m = 2.82 μm is the size of the largest particles that can still remain in suspension.

2.5 Sedimentation and Centrifugation

If an object is not spherical, we can still use the basic approach just described to determine its terminal velocity. The only difference will be that we shall not be able to use Stoke's law for the resistive force. Instead, let us assume that the resistive force has a magnitude given by $F_r = kv$, where k is a coefficient of frictional resistance that must be determined experimentally. As we discussed above, the object reaches its terminal speed when the weight downward is balanced by the net upward force, or

$$w = B + F_r \tag{2.10}$$

where B is the buoyant force, given by $B = \rho_f g V$.

We can use the fact that the volume, V, of the displaced fluid is related to the density of the falling object, ρ, by $V = m/\rho$. Hence, we can express the buoyant force as

$$B = \frac{\rho_f}{\rho} mg$$

Let us substitute this expression for B and $F_r = kv_t$ into Equation 2.10 (terminal speed condition):

$$mg = \frac{\rho_f}{\rho} mg + kv_t$$

or

$$v_t = \frac{mg}{k}\left(1 - \frac{\rho_f}{\rho}\right) \tag{2.11}$$

The terminal speed for particles in biological samples is usually quite small. For example, the terminal speed for blood cells falling through plasma is about 5 cm/h in the gravitational field of the Earth. The terminal speeds for the molecules that make up a cell are many orders of magnitude smaller than this because of their much smaller mass. The speed at which materials fall through a fluid is called the *sedimentation rate*. This number is often important in clinical analysis.

It is often desired to increase the sedimentation rate in a fluid. A common method used to accomplish this is to increase the effective acceleration g that appears in Equation 2.11. A fluid containing various biological molecules is placed in a centrifuge and whirled at very high angular velocities (Fig. 2.12). Under these conditions, the particles experience a large radial acceleration, $a_c = v^2/r = \omega^2 r$, which is much greater than the acceleration due to gravity, and so we can replace g in Equation 2.11 by $\omega^2 r$:

$$v_t = \frac{m\omega^2 r}{k}\left(1 - \frac{\rho_f}{\rho}\right) \tag{2.12}$$

This equation indicates that those particles having the greatest mass will have the largest terminal speed. Therefore, the most massive particles will settle out on the bottom of a test tube first.

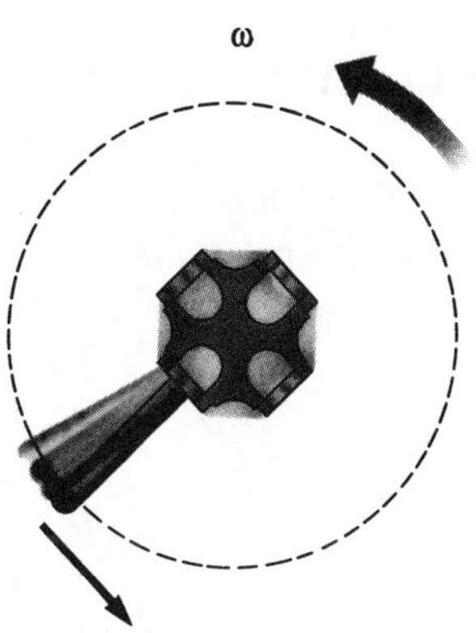

Figure 2.12
Simplified diagram of a centrifuge (top view).

Example 2.6 The Spinning Test Tube

A centrifuge rotates at 50,000 rev/min, which corresponds to an angular frequency of 5240 rad/s (a typical speed). A test tube placed in this device has its top 5 cm from the axis of rotation and its bottom 13 cm from this axis. Find the effective value of g at the midpoint of the test tube, which corresponds to a distance 9 cm from the axis of rotation.

Solution
The acceleration experienced by the particles of the tube at a distance r=9 cm from the axis of rotation is given by

$$a_c = \omega^2 r = \left(5240 \frac{\text{rad}}{\text{s}}\right)^2 (9 \times 10^{-2}\ \text{m}) = 2.47 \times 10^6\ \text{m/s}^2$$

Exercise If the mass of the contents of the test tube is 15 g, find the centripetal force that the bottom of the tube must exert on the contents of the tube. Assume a centripetal acceleration equal to that found at the midpoint of the tube.
Answer 3.71×10^4 N, or about 8000 lb! (Because of such large forces, the base of the tube in a centrifuge must be rigidly supported to keep the glass from shattering.)

ESSAY

Physics of the Human Circulatory System

William G. Buckman
Western Kentucky University

The human circulatory system is an extremely complex and vital part of the human body. The blood supplies food and oxygen to the tissues of the body, carries away the waste products from the cells, distributes the heat generated by the cells to equalize the temperature of the body, carries hormones that stimulate and coordinate the activity of organs, distributes antibodies to fight infection, and performs numerous other functions.

William Harvey (1579-1657), an English physician and physiologist, studied blood flow and the action of the heart. He established the essential mechanics of the heart and found that the blood flows from the arterial system through capillary beds and into the veins to be returned to the heart.

The Physical Properties of Blood

Blood is a liquid tissue consisting of two principal parts: the plasma, which is the intercellular fluid, and the cells, which are suspended in the plasma. Plasma is about 90% water, 9% proteins, and 0.9 % salts, sugar, and traces of other materials. Blood contains white blood cells and red blood cells. The individual red blood cells are biconcave and have an average diameter of 7.5 μm. There are about 5×10^6 red blood cells per cubic millimeter of blood. The five types of white blood cells found in the blood have an average concentration of 8000 per cubic millimeter, with the concentration normally varying between 4500 and 11,000 per cubic millimeter. The density of blood is about 1.05×10^3 kg/m^3, and its viscosity varies from 2.5 to 4 times that of water.

The Heart as a Pump

The heart can be considered as a double pump, with each side consisting of an atrium and a ventricle (Figure 1a). Blood enters the right atrium, flows into the right ventricle, is pumped by the right ventricle. The left ventricle then pumps the oxygenated blood out through the aorta to the rest of the body. The heart has a system of one-way valves to assure that the blood flows in the proper direction. The heart's pumping cycle has the two ventricles pumping at the same time, as shown in Figure 1b.

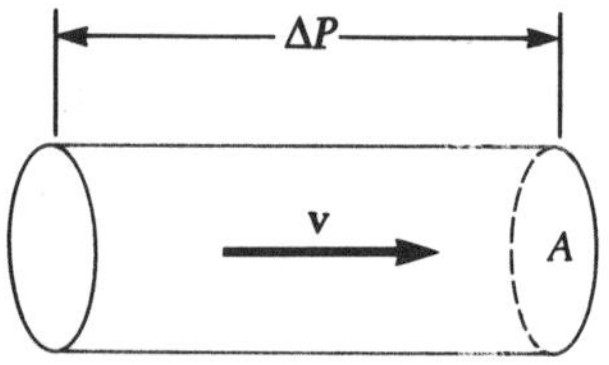

Figure 2
The power required to maintain blood flow against viscous forces.

The pressure generated by the right ventricle is quite low (about 25 mm Hg), and the lungs offer a low resistance to blood flow. The left ventricle generates a larger pressure, typically greater than 120 mm Hg, at the peak (systole) of the pressure. During the resting stage (diastole) of the heartbeat, the pressure is typically about 80 mm Hg.

We shall now calculate the mechanical work done by the heart. Consider the fluid in the vessel shown in Figure 2. The net force on the fluid is equal to the product of the pressure drop across the fluid, ΔP, and the cross-sectional area, A. The power expended is equal to the net force times the average velocity: $(\Delta PA)(\bar{v})$. Because $A = AL/t$= volume / time, which is the flow rate, we may now write for the power expended by the heart

$$\text{Power} = (\text{flow rate})(\Delta P)$$

If a normal heart pumps blood at the rate of 97 cm^3/s and the pressure drop from the arterial system to the venous system is 1.17×10^4 Pa, we then have

$$\text{Power} = (97\ \text{cm}^3/\text{s})(10^6\ \text{m}^3/\text{cm}^3)(1.17 \times 10^4\ \text{Pa}) = 1.1\ \text{W}$$

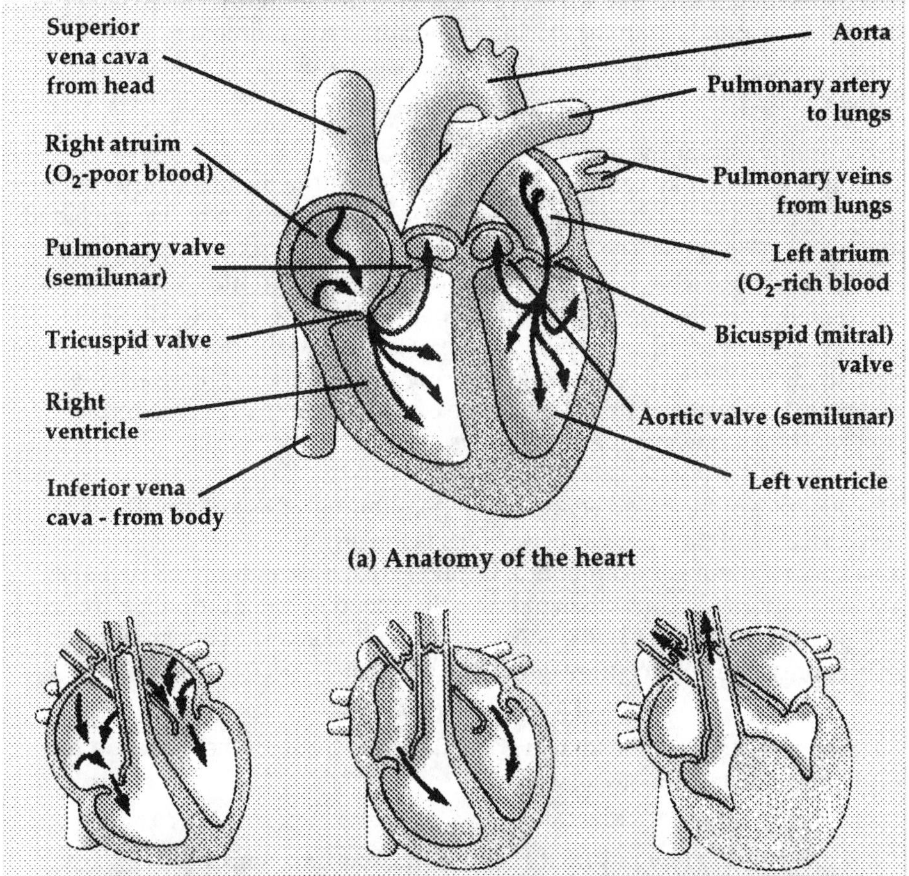

1. Blood fills both atria, some blood flows into ventricles - diastole phase of artria.
2. Atria contract, squeezing blood into ventricles - ventricular diastole.
3. Ventricles contract, squeezing blood into aorta and pulmonary arteries - ventricular systole phase.

(b) Pumping cycle of the heart

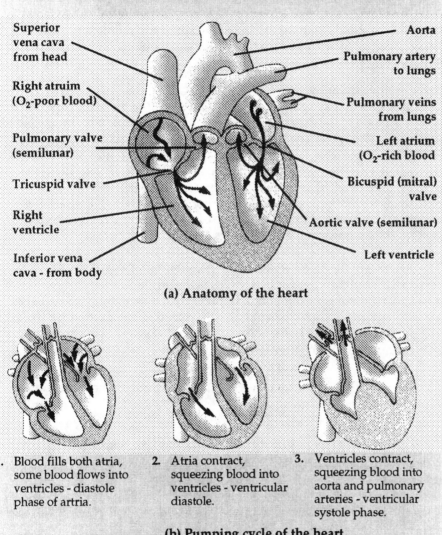

Figure 1
(a) Anatomy of the heart. (b)Pumping cycle of the heart.

By measuring oxygen consumption, it is found that the heart of a 70-kg man at rest consumes about 10 W. In the calculation above, it was determined that 1.1 W is required to do the mechanical work of pumping blood; hence, the heart is typically about 10% efficient. During strenuous exercise, the blood pressure may increase by 50% and the blood volume pumped may increase by a factor of 5 to yield an increase of 7.5 times in the power generated by the left ventricle. Because the right ventricle has a systolic pressure about one fifth that of the left ventricle, its power requirement is about one fifth that of the left ventricle.

When we listen to a heart with a stethoscope, we hear two sharp sounds. The first corresponds with the closing of the tricuspid and mitral valves, and the second corresponds with the closing of the aortic and pulmonary valves. Other sounds that are heard are those associated with the flow and turbulence of the blood.

The Cardiovascular System

The cardiovascular system includes the heart to pump the blood; arteries to carry the blood to the organs, muscle, and skin; and veins to return the blood to the heart (Figure 3). The blood is pumped into the aorta by the left ventricle. The aorta branches to form smaller arteries, which in turn branch down to even smaller arteries, until finally the blood reaches the very small capillaries of the vascular bed. These capillaries are so small that the red blood cells must pass single file through them. After passing through the capillaries, where materials being carried by the blood are exchanged with the surrounding tissues, the blood flows to the veins and is returned to the heart.

The flow rate of the blood changes as it goes through this system. The cross-sectional area of the vascular bed, which is the product of the cross-sectional area and the number of capillaries, is much greater than the cross-sectional area of the aorta. Because the volume of the blood passing through a cross-sectional area per unit of time is Av, where v is the speed of the blood, we may express the volume flow rate of the blood as

$$\text{Flow rate} = A_{aorta}\, v_{aorta}$$

Furthermore, because the total average flow rate through the aorta and the capillaries must be the same, we have

$$\text{Flow rate} = A_{aorta}\, v_{aorta} = A_{cap}\, v_{cap}$$

Example 10E.1 Flow of Blood in the Aorta and Capillaries

The speed of blood in the aorta is 50 cm/s, and it has a radius of 1 cm.
(a) What is the rate of flow of blood through this aorta? (b) If the capillaries have a total cross-sectional area of 3000 cm^2, what is the speed of the blood in the capillaries?

Solution

(a) The area of the aorta is

$$A = \pi r^2 = \pi(1\text{ cm})^2 = \pi\text{ cm}^2$$

$$\text{Flow rate} = Av = (\pi\text{ cm}^2)(50\text{ cm/s}) = 50\pi\text{ cm}^3/\text{s}$$

(b) The flow rate in the capillaries = $50\,\pi$ cm^3/s $= A_c v_c$,

$$v_c = \frac{\text{Flow rate}}{A_c} = \frac{50\pi\ \text{cm}^3/\text{s}}{3000\ \text{cm}^2} = 0.05\ \text{cm/s}$$

This low blood speed in the capillaries is necessary to enable the blood to exchange oxygen, carbon dioxide, and other nutrients with the surrounding tissues.

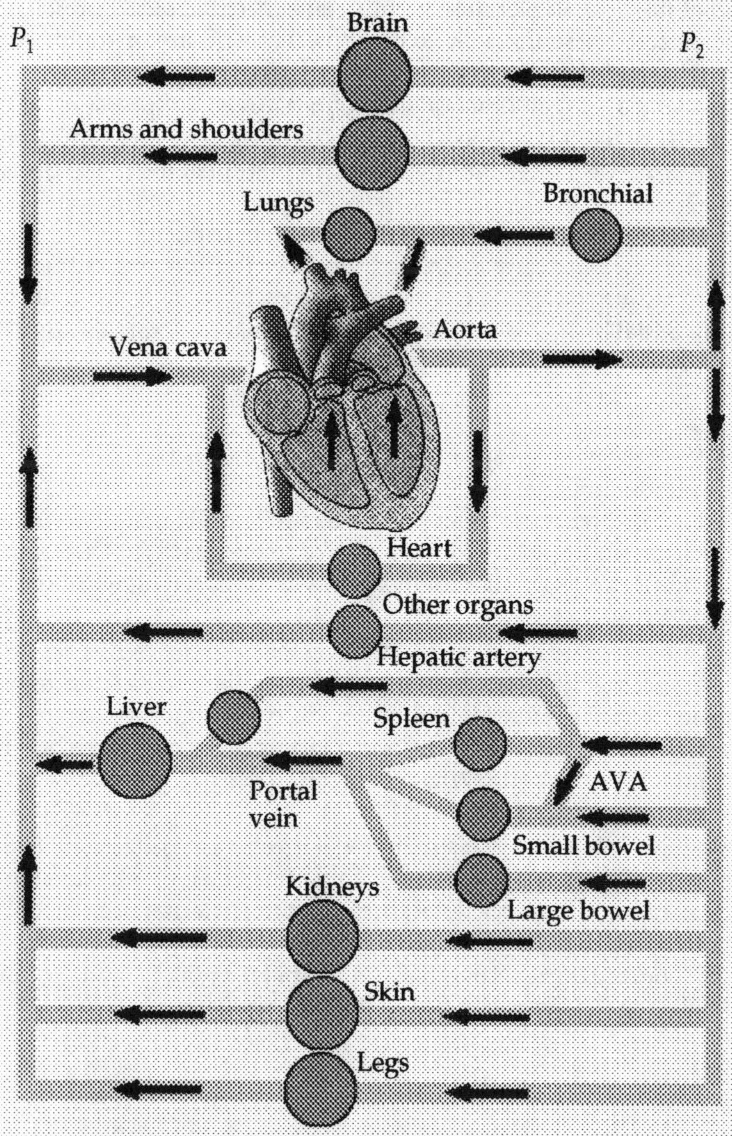

Figure 3
A diagram of the mammalian circulatory system. Pressure P_2 is that in the arterial system, and P_1 is that in the venous system. Arrows indicate the direction blood flow.

Essay Questions

1. Explain why some individuals tend to black out when they stand up rapidly.
2. If a person is standing and is at rest, what will be the relation between the blood pressure in the left arm and the left leg? What is the relation if the person is in a horizontal position?
3. At what upward acceleration would you expect the blood pressure in the brain to be zero? (Assume that no body mechanisms are operating to compensate for this condition.)
4. When using a sphygmomanometer to measure blood pressure, will the blood pressure readings depend upon the atmospheric pressure? If the atmospheric pressure decreases rapidly, how will this effect the blood pressure readings?
5. Why is it impractical to measure the pulse rate using a vein?
6. Assuming that an artery is clogged such that the effective radius is one-half its normal radius compared to a normal artery, by what factor must the pressure differential be increased to obtain the normal flow rate through the clogged artery?

Problems

1. Determine the average speed of the blood in the aorta if it has a radius of 1.2 cm, and the flow rate is 20 liters/min.
2. If the mean blood pressure in the aorta is 100 mm Hg, (a) determine the blood pressure in the artery located 2 ft above the heart. (b) One cannot apply, without significant error, Bernoulli's principle in the smaller arteries and the capillaries. Why not?
3. When the flow rate is 5 liters/min, the velocity in the capillaries is 0.33 mm/s. Assuming the average diameter of a capillary to be 0.008 mm, calculate the number of capillaries in the circulatory system.
4. An artery has a length of 20 cm, and a radius of 0.5 cm, and blood is flowing at a rate of 6 liters/min. What is the difference in the pressure between the ends of the artery?
5. Assuming the internal pressure in the left ventricle is 100 mm Hg and the left ventricle of the heart has an effective radius of 3 cm, calculate the tension in the wall of the left ventricle. (Assume that the external pressure is zero gauge pressure).
6. If the blood pressure is 10 mm Hg in a capillary which has a radius of 0.005 mm, determine the tension in the wall of the capillary.

Section 3
ACOUSTICS

3.1 The Ear

The human ear is divided into three regions; the outer ear, the middle ear, and the inner ear (Figure 3.1a). The outer ear consists of the ear canal (open to the atmosphere), which terminates on the eardrum (tympanum). Sound waves travel down the ear canal to the eardrum, which vibrates in and out in phase with the pushes and pulls caused by the alternating high and low pressures of the sound wave. Behind the eardrum are three small bones of the middle ear, called the hammer, the anvil, and the stirrup because of their shapes (Figure 3.1b). These bones transmit the vibration to the inner ear, which contains the cochlea, a snail-shaped tube about 2 cm long. The cochlea makes contact with the stirrup of the oval window and is divided along its length by the basilar membrane, which consists of small hairs and nerve fibers. This membrane varies in mass per unit length and tension along its length, and different portions of it resonate at different frequencies. (Recall that the natural frequency of a string depends on its mass per unit length and on the tension on it.) Along the basilar membrane are numerous nerve endings, which sense the vibration of the membrane and in turn transmit impulses to the brain. The brain interprets the impulses as sounds of varying frequency, depending on the location along the basilar membrane of the impulse-transmitting nerves and on the rate at which the impulses are transmitted.

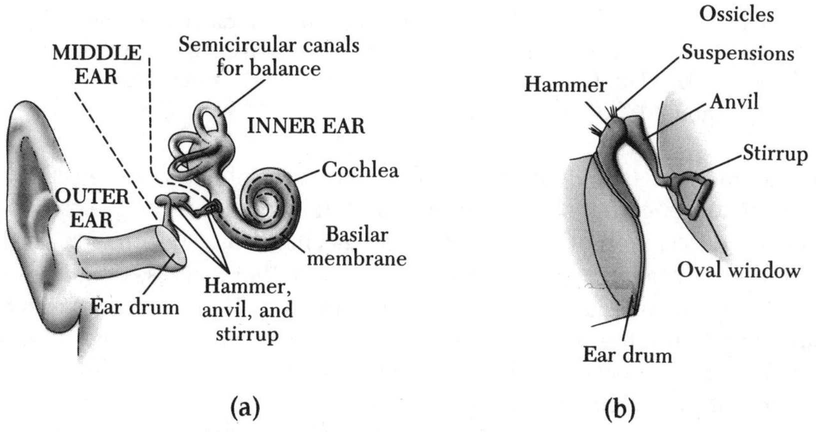

Figure 3.1
(a) Structure of the human ear. (b) The three tiny bones (ossicles) that connect the eardrum to the window of the cochlea act as a double lever system to decrease the amplitude of vibration and hence increase the pressure on the fluid in the cochlea.

Figure 3.2 is a frequency response curve for an average human ear for sounds of equal loudness, ranging from 0 dB to 120 dB. It shows that the decibel level at which pain is experienced does not vary greatly from 120 dB regardless of the frequency of the tone. However, the threshold of hearing shown by the lowest curve is very strongly dependent on frequency. The easiest frequencies to hear are around 3300 Hz, whereas frequencies about 12,000 Hz or below about 50 Hz must be relatively intense to be heard. A person with very good hearing (about 1% of the population) would have a threshold of hearing corresponding to the lowest curve. About 50% of the population has a threshold of hearing corresponding to the second lowest curve.

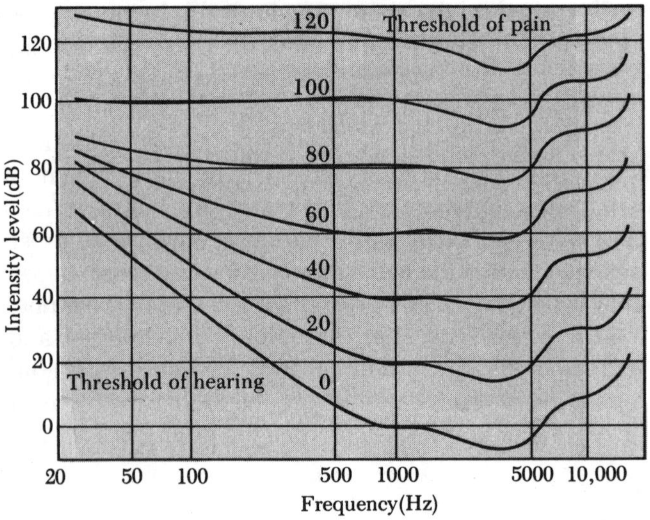

Figure 3.2

Curves of intensity level versus frequency for sounds that are perceived to be of equal loudness. Note that the ear is most sensitive at a frequency of about 3300 Hz. The lowest curve corresponds to the threshold of hearing for only about 1% of the population.

The exact mechanism by which sound waves are amplified and detected by the ear are rather complex and not fully understood. However, we can give a qualitative description of the amplification mechanisms. The small bones in the middle ear represent an intricate lever system that increases the force on the oval window over a given force on the eardrum. The pressure is greatly magnified because the surface area of the eardrum is about 20 times that of the oval window (in analogy with the hydraulic press). The middle ear, together with the eardrum and oval window, in effect acts as a matching network between the air in the outer ear and the liquid in the inner ear. the overall energy transfer between the outer ear and inner ear is highly efficient, with pressure amplification factors of several thousand. In other words, pressure variations in the inner ear are much greater than those in the outer ear. The ear has its own built-in protection against loud sounds. The muscles connecting the three bones to the walls of the middle ear control the volume of the sound by changing the tension on the bones as sound builds up, thus hindering their ability to transmit vibrations. In addition, the eardrum becomes stiffer. These two occurrences cause the ear to be less sensitive to loud sounds. There is a time delay between the onset of loud sound and the ear's protective reaction, however, so that a very sudden loud sound can still damage the ear.

3.2 Ultrasound and Its Applications

Ultrasonic waves are sound waves whose frequencies are in the range of 20 kHz to 100 kHz, which is beyond the audible range. Because of their high frequency, and corresponding short wavelengths, ultrasonic waves can be used to produce images of small objects and are currently in wide use in medical applications, both as a diagnostic tool and in certain treatments. Various internal organs in the body can be examined through the images produced by the reflection and absorption of ultrasonic waves. Although ultrasonic waves are far safer than x-rays, their images do not always provide as much detail. On the other hand, certain organs, such as the liver and the spleen, are invisible to x-rays but can be diagnosed with ultrasonic waves.

It is possible to measure the speed of blood flow in the body using a device called an ultrasonic flow meter. The technique makes use of the Doppler effect. By comparing the frequency of the waves scattered by the blood vessels with the incident frequency, one can obtain the speed.

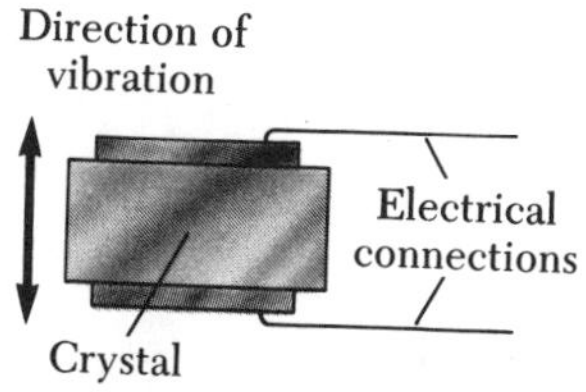

Figure 3.3
An alternating voltage applied to the faces of a piezoelectric crystal causes the crystal to vibrate.

The technique used to produce ultrasonic waves for clinical use is illustrated in Figure 3.3. Electrical contacts are made to the opposite faces of a crystal, such as quartz or strontium titanate. If an alternating voltage of very high frequency is applied to these contacts, the crystal will vibrate at the same frequency as the applied voltage. As the crystal vibrates, it emits a beam of ultrasonic waves. At one time, almost all of the headphones used in radio reception produced their sound in this manner. This method of transforming electrical energy into mechanical energy is called the piezoelectric effect. This effect is also reversible. That is, if some external source causes the crystal to vibrate, an alternating voltage is produced across the crystal. Hence, a single crystal can be used to both transmit and receive ultrasonic waves.

The production of electric voltages by a vibrating crystal is a technique that has been used for years in stereo and hi-fi equipment. In this application, a phonograph needle is attached to the crystal, and the vibrations of the needle as it rides in the groove of the record are translated by the crystal into an alternating voltage. This voltage is then amplified and used to drive the system's speakers.

The primary physical principle that makes ultrasound imaging possible is the fact that a sound wave is partially reflected whenever it is incident on a boundary between two materials having different densities. It is found that, if a sound wave is traveling in a material of density ρ_i and strikes a material of density ρ_t, the percentage of the incident sound wave reflected, *PR*, is given by

$$PR = \left(\frac{\rho_i - \rho_t}{\rho_i + \rho_t} \right)^2 \times 100 \qquad \textbf{(3.1)}$$

This equation assumes that the incident sound wave travels perpendicular to the boundary and that the speed of sound is approximately the same in both materials. This latter assumption holds very well for the human body since the speed of sound does not vary much in the various organs of the body.

Physicians commonly use ultrasonic waves to observe a fetus. This technique offers far less risk than x-rays, which can be genetically dangerous to the fetus and can produce birth defects. First the abdomen of the mother is coated with a liquid, such as mineral oil. If this is not done, most of the incident ultrasonic waves from the piezoelectric source will be reflected at the boundary between the air and the skin of the mother. Mineral oil has a density similar to that of skin, and as Equation 3.1 indicates, a very small fraction of the incident ultrasonic wave is reflected when $\rho_i \approx \rho_t$. The ultrasound energy is emitted as pulses rather than as a continuous wave so that the same crystal can be used as a detector as well as a transmitter. The source-receiver is then passed over a particular line along the mother's abdomen. The reflected sound waves picked up by the receiver are converted to an electric signal, which forms an image along a line on a fluorescent screen. The sound source is then moved a few centimeters on the mother's body, and the process is repeated. The reflected signal produces a second line on the fluorescent screen. In this fashion a complete scan of the fetus can be made. Difficulties with the pregnancy, such as the likelihood of abortion or of breech birth, are easily detected with this technique. Also, such fetal abnormalities as spina bifida and water on the brain are readily observable.

A relatively new medical application of ultrasonics is the **CUSA**, for cavitron ultrasonic surgical aspirator. These devices have made it possible to surgically remove brain tumors that were previously inoperable. The cusa is a long needle that emits very high frequency ultrasonic waves, about 23 kHz, at its tip. When the tip touches a tumor, the part of the tumor near the needle is shattered and the residue can be sucked up (aspirated) through the hollow needle.

Another interesting application of ultrasound is the ultrasonic ranging unit designed by the Polaroid Corporation. This device is used in some of their cameras to provide an almost instantaneous measurement of the distance between the camera and object to be photographed. The principal component of this device is a crystal that acts as both a loudspeaker and a microphone. A pulse of ultrasonic waves is transmitted from the transducer to the object to be photographed. The object reflects part of the signal, producing an echo that is detected by the device. The time interval between the outgoing pulse and the detected echo is then electronically converted to a distance value, since the speed of sound is a known quantity.

3.3 Audiometry

Audiometry is a common medical procedure used to test for hearing loss. In this technique, sound signals of known intensity and frequency are introduced through a headset into one ear of the patient. The frequencies normally used are 125, 250, 500, 750, 1000, 2000, 3000, 4000, 6000, and 8000 Hz. The results of the test are usually plotted on a graph like the one shown in Figure 3.4. The vertical axis represents the decibel level that the sound must be raised above threshold to be heard, and the horizontal axis is the frequency of the sound. The results illustrated in figure 3.4 indicate some hearing loss from 3000 Hz upward.

When hearing loss is severe in one ear, there are difficulties with the hearing test because sound waves are transmitted through the bones of the head to the normal ear. In order to overcome this difficulty, a technique called masking is often used. The principle behind the masking procedure is to apply a masking sound signal to the normal ear so that it will be occupied by this stimulus while the test is being carried out on the impaired ear. This inhibits the crossover effect and produces more reliable results.

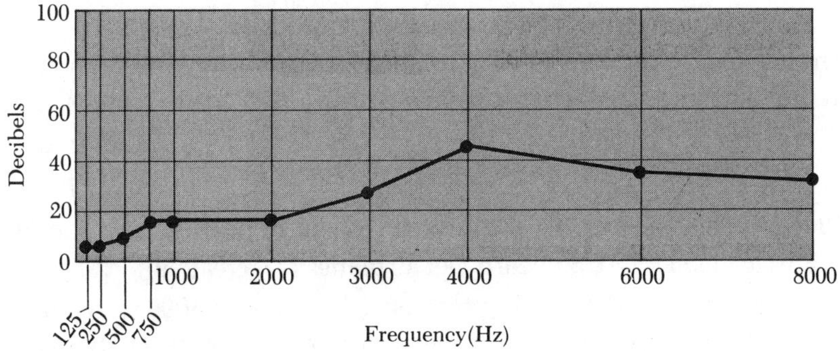

Figure 3.4
Hearing analysis by an audiometer.

3.4 Intensity Levels in Decibels

The faintest sounds the human ear can detect at a frequency of 1000 Hz have an intensity of about 10^{-12} W/m^2. This intensity is called the **threshold of hearing**. The loudest sounds the ear can tolerate have an intensity of about 1 W/m^2 (**the threshold of pain**). At the threshold of hearing, the increase in pressure in the ear is approximately 3×10^{-5} Pa over normal atmospheric pressure. Since atmospheric pressure is about 10^5 Pa, this means the ear can detect pressure fluctuations as small as about 3 parts in 10^{10}! Also, at the threshold of hearing, the maximum displacement of an air molecule is about 1×10^{-11} m. This is a remarkably small number! If we compare this result with the diameter of a molecule (about 10^{-10} m) , we see that the ear is an extremely sensitive detector of sound waves.

In a similar manner, one finds that the loudest sounds the human ear can tolerate correspond to a pressure increase of about 29 Pa over normal atmospheric pressure and a maximum displacement of air molecules of 1×10^{-5}m.

Thus, the human ear can detect a wide range of intensities, with the loudest tolerable sounds having intensities about 10^{12} times greater than those of the faintest detectable sounds. However, the loudness of the most intense sound is not preceived as being 10^{12} times greater than that of the faintest sound. This is because the sensation of loudness is approximately logarithmic in the human ear. The relative loudness of a sound is called the **intensity level** or **decibel level,** β. This unit is named after the inventor of the telephone, Alexander Graham Bell (1847-1922) and is defined as

$$\beta = 10 \log\left(\frac{I}{I_0}\right) \qquad \textbf{(3.2)}$$

The constant I_0 is the reference intensity level, taken to be the sound intensity at the threshold of hearing ($I_0 = 10^{-12}$ **W/m^2**), and I is the intensity at the level β, where β is measure in decibel (dB). On this scale, the threshold of pain (I = 1W/m^2) corresponds to an intensity level of $\beta = 10 \log (1/10^{-12}) = 10 \log(10^{12}) = 120$ dB. Likewise, the threshold of hearing corresponds to $\beta = 10 \log (1/1) = 0$ dB. Nearby jet airplanes can create intensity levels of 150 dB, and subways and riveting machines have levels of 90 to 100 dB. The electronically amplified sound heard at rock concerts can be at levels of up to 120 dB, the threshold of pain. Prolonged exposure to such high intensity levels can produce serious damage to the ear. Ear plugs are recommended whenever intensity levels exceed 90 dB. Recent evidence suggests that noise pollution, which is common in most large cities and in some industrial environments, may be a contributing factor tohigh blood pressure, anxiety, and nervousness. Table 3.1 gives some idea of the intensity levels of various sounds.

Table 3.1 Intensity Levels in Decibels for Some Sources

Source of Sound	β(dB)
Nearby jet airplane	150
Jackhammer, machine gun	130
Siren, rock concert	120
Subway, power mower	100
Busy traffic	80
Vacuum cleaner	70
Normal conversation	50
Mosquito buzzing	40
Whisper	30
Rustling leaves	10
Threshold of hearing	0

Example 3.1 Intensity Levels of Sound

Calculate the intensity level of a sound wave having an intensity of (a) 10^{-12} W/m^2, (b) 10^{-11} W/m^2, and (c) 10^{-10} W/m^2.

Solution

(a) For an intensity level of 10^{-12} W/m^2, the intensity level in decibels is

$$\beta = 10 \log\left(\frac{10^{-12}\,W/m^2}{10^{-12}\,W/m^2}\right) = 10 \log(1) = 0 \text{ dB}$$

This answer should have been obvious without calculation because an intensity of 10^{-12} W/m^2 corresponds to the threshold of hearing.

(b) In this case, the intensity is exactly 10 times greater than in Part (a). The intensity level is

$$\beta = 10 \log\left(\frac{10^{-11}\,W/m^2}{10^{-12}\,W/m^2}\right) = 10 \log(10) = 10 \text{ dB}$$

(c) Here the intensity is 100 times greater than the intensity at the threshold of hearing and the intensity level is

$$\beta = 10 \log\left(\frac{10^{-10}\,W/m^2}{10^{-12}\,W/m^2}\right) = 10 \log(100) = 20 \text{ dB}$$

Note the pattern in these answers. A sound with an intensity level of 10-dB is 10 times more intense than the 0-dB sound, and a sound with an intensity level of 20-dB is 100 times more intense than a 0-dB sound. This pattern is continued throughout the decibel scale. In short, on the decibel scale *an increase of 10-dB means that the intensity of the sound increases by a factor of 10.* For example, a 50-dB sound is 10 times more intense than a 40-dB sound and a 60-dB sound is 100 times more intense than the 40-dB level.

PROBLEM

ACOUSTICS

1. A rather noisy typewriter is found to have a sound intensity of 10-5 W/m^2. Find the decibel level of this machine and calculate the new decibel level when a second identical typewriter is added to the office.

SOLUTION

ACOUSTICS

1. The decibel level of the single typewriter is

$$\beta = 10 \log\left(\frac{10^{-5}\,\mathrm{W/m^2}}{10^{-12}\,\mathrm{W/m^2}}\right) = 10 \log(10^7) = 70\ \mathrm{dB}$$

Adding the second typewriter doubles the energy input into sound and hence doubles the intensity. The new decibel level is

$$\beta = 10 \log\left(\frac{2 \times 10^{-5}\,\mathrm{W/m^2}}{10^{-12}\,\mathrm{W/m^2}}\right) = 73\ \mathrm{dB}$$

Federal regulations now demand that no office or factory worker can be exposed to noise levels that average more than 90 dB over an 8 hour day. The results in this example illustrate some of the problems associated with noise abatement. First, note that when the second typewriter was added, the noise level increased by only 3 dB. Because of the logarithmic nature of decibel levels, doubling the intensity does not double the decibel level; in fact, it alters it by a surprisingly small amount. This means that additional equipment can be added to an office or factory without appreciably altering the decibel level of an environment.

The down side to this observation is that the results also work in reverse. As you remove noisy machinery, the decibel level is not lowered appreciably. For example, consider an office with 60 typewriters producing a noise level of 93 dB, which is 3 dB above the maximum allowed. In order to reduce the noise level by 3 dB, half the machines would have to be removed! That is, you would have to remove 30 typewriters to reduce the noise level to 90 dB. To reduce the level another 3 dB, one would have to remove half the remaining machines, and so on.

Section 4
BASAL METABOLIC RATE

Any living thing is a heat producer. A working muscle such as the heart or limbs produces heat. The metabolic process itself results in a release of energy in the form of heat. All this continuous heat production would result in a steady temperature rise if the heat was not carried away. A warm object does lose heat to its surroundings; the greater the temperature difference, the faster the heat is lost. If energy is produced at a constant rate in an object, the temperature rises until the rate of loss just balances the rate of production, as in Figure 4.1. An equilibrium condition results.

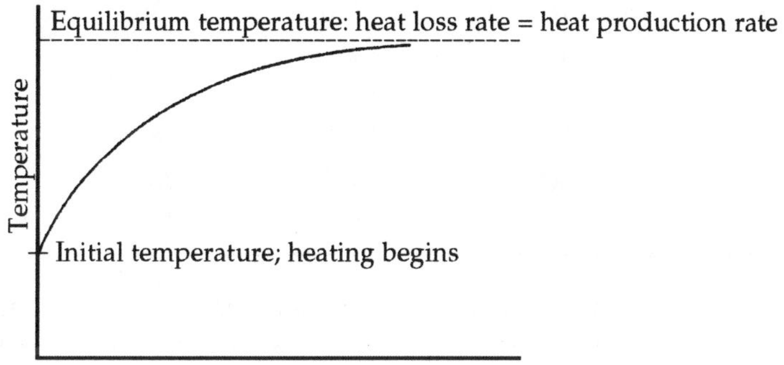

Figure 4.1
The temperature of an object with an internal heat source.

A so-called cold-blooded creature, such as a reptile or fish, will be a few degrees warmer than its surroundings. A warm-blooded creature, on the other hand, keeps its body temperature constant. The rate of heat dissipation must also be comparatively constant, increasing somewhat with increased heat production as in exercise; but the rate of heat loss must be made almost independent of the surrounding temperature. Our bodies, even while resting, must dissipate about 100 watts, no matter whether our surroundings are cool or warm, whether it is a cold winter day or a hot summer day.

A large person uses more energy and hence produces more heat than does a small one; a cow produces more heat than a mouse, and so on. To compare people or animals of different size, we simplify the problem by working with the energy production in the resting state – the basal metabolic rate or BMR. Consider size to be measured by mass. The effect of change in mass is most apparent if different animals are considered; the data in Table 4.1 show the basal metabolic rates for various animals ranging from a dove to a steer. The units are kcal/day and watts. It is not surprising that the BMR rises with increasing mass, but the last column of Table 4.1 is interesting. The heat production per unit mass drops from 125 cal/g each day for a dove to only 13 cal/g each day for a steer. Figure 4.2 is a graph of the data in columns 2 and 3 of Table 4.1. If the metabolic rate varied directly as the mass, Figure 4.2 would be a straight line. It is not

Table 4.1 Basal metabolic rates of a few animals.

	MASS, *M*	BASAL METABOLIC RATE, *E/t*		
ANIMAL	KG	*kcal/day*	*watts*	*cal/gm day*
dove	0.16	20	0.97	125
rate	0.26	30	1.45	115
pigeon	0.30	32	1.55	107
hen	2.0	100	4.8	50
dog(female)	11	300	14.5	27
dog(male)	16	420	20	26
sheep	45	1050	50	23
woman	60	1400	68	23
man	70	1800	87	26
cow	400	5500	266	14
steer	680	8500	411	13

straight, but all of the points do follow along a curve; they are not random. This shows that there is a relation between metabolic rate and mass. (a) What is the relation between BMR and mass, or, in other words, the equation relating them?

(b) Why does the relation take the form that it does? In answering these questions, a measure of understanding of the processes involved may be obtained.

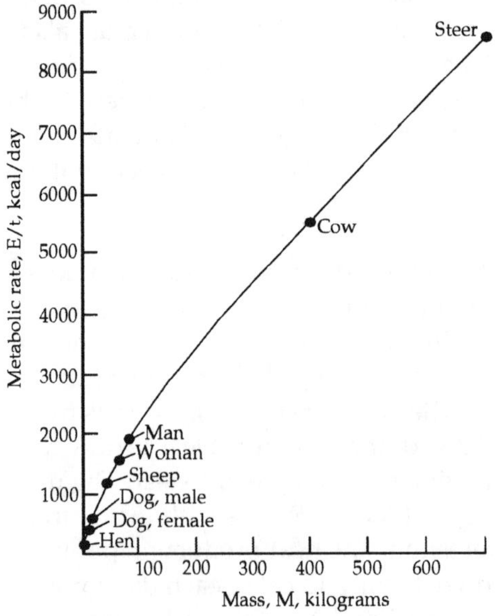

Figure 4.2
Basal metabolic rates of various animals, plotted against body mass.

In tackling this problem, some thought analysis about the factors involved will give some hints about what to expect.

To a first approximation, all of the animals involved can be considered to be about the same number of degrees warmer than the surroundings. A limitation like this must be remembered later. Most of the heat loss takes place through the surface,* and for a given temperature difference above the surroundings and a given type of surface (another limitation on the analysis) each unit area loses heat at the same rate, be it a steer, man, or pigeon. Since heat production equals heat loss, perhaps the heat production is proportional to surface area. The term *proportional* to means that the quantities are related by an equation of the form

$$\text{BMR} = kA$$

where BMR is the basal metabolic rate, the rate of heat production and of heat loss. It will perhaps be more meaningful to let BMR be represented by E/t, where E is a quantity of energy (heat) lost or produced in a time t so that E/t is a rate of energy transfer.

The surface area is A, and k is a proportionality constant.
So perhaps

$$\frac{E}{t} = kA$$

Now, surface area does depend on the mass for objects of similar proportions and densities. The mass of living objects is proportional to their volume. The volume is proportional to the *cube* of the linear dimensions, while the area is proportional to the *square* of the linear dimensions. For example, the volume of a sphere is given by

$$V = \frac{4}{3}\pi R^3$$

while its surface area is given by

$$V = 4\pi R^2$$

There are similar relations for other shapes. In symbols, if L is a linear dimension and the mass M is directly proportional to volume, then

$$M \propto L^3$$

or

$$L \propto \sqrt[3]{M} \text{ or } L \propto M^{1/3}$$

That is, if the mass is proportional to the cube of a linear dimension, then the linear dimension is proportional to the cube root of the mass. Also

$$A \propto L^2$$

* This treatment neglects respiratory heat loss, which is quite important in some animals.

so $$A \propto (M^{1/3})^2$$

or $$A = k' M^{2/3}$$

where k' is another constant. If it is expected that perhaps

$$\frac{E}{t} = kA$$

then the expected relation to M is

$$\frac{E}{t} = k'' M^{2/3}$$

$$= k'' M^{0.667}$$

Here k'' is just another constant, but the suggestion is that metabolic rate, BMR or E/t, varies as the 2/3 power of the mass. If it does, then we do understand something about the heating process in a body.

The next step is to find how E/t is *actually* related to M, using the measured data from Table 8-8. It is not to try to prove that E/t is related to $M^{2/3}$. In trying to *prove* something, there is an inclination to think it has been proven when it really hasn't. It is better to develop the theoretical relation and to find the real relation as independently as possible. They will not be expected to be the same because of the approximations assumed in the analysis. If the two relations are reasonably close, then it is said that perhaps the analysis is on the right track and that the discrepancies result from the limitations put on the situation analyzed: in this example, the same temperature excess, same surface type, and same relative shape.

The next step in this example is to use the given data about BMR (or E/t) and M to find the equation that describes those data. The theoretical analysis suggested that perhaps the relation is of a power form. That is, perhaps

$$\left(\frac{E}{t}\right) = kM^p$$

k is just another constant

p is a power, perhaps near 0.667

A suspected power relation can be analyzed graphically by first taking the logs of both sides of the equation. Remember, too, that

$$\log ab = \log a + \log b$$

and $$\log b^c = c \log b$$

Then $$\log E/t = \log k + p \log M$$

To graph this, let:

$y = \log (E/t)$
$x = \log M$
$a = \log k$ (this is the y intercept, y_0)

The equation being graphed is then of the form

$$y = a + px$$

This is the equation of a straight line whose slope is the power p. The data for the graph are shown in Table 4.2, and the graph itself is in Figure 4.3. The best fit to the data is indeed a straight line, so metabolic rate *is* related to mass by a power law. The power p to which M is raised is found from a slope triangle, as shown in Figure 4.3. The slope in this case is 0.72. The y intercept is not so important, but it is 1.85. The constant k is the antilog of this, or 71.

The equation relating basic metabolic rate and mass is then

$$\frac{E}{t} = 71\, M^{0.72}$$

with E/t in kcal/day and M in kilograms. This is not quite what was theoretically expected. The next question, "Why the discrepancy?" will be left to provoke thought. However, the process of science, analysis of how two things are really related and a theoretical analysis of how they are expected to be related, has been demonstrated. If there is reasonable correlation, we feel that we have some understanding of what is going on.

Table 4.2 The BMR and mass from Table 4.1, and the logarithms of these quantities. These data are used to plot Figure 4.3.

Animal	*E/t*, kcal/day	M, kg	LOG (*E/t*)	LOG *M*
dove	20	0.16	1.30	-0.80
rat	30	0.26	1.48	-0.59
pigeon	32	0.30	1.51	-0.52
hen	100	2.0	2.00	0.30
dog (female)	300	11	2.48	1.04
dog (male)	420	16	2.62	1.20
sheep	1050	45	3.02	1.65
woman	1400	60	3.15	1.78
man	1800	70	3.26	1.85
cow	5500	400	3.74	2.60
steer	8500	680	3.93	2.83

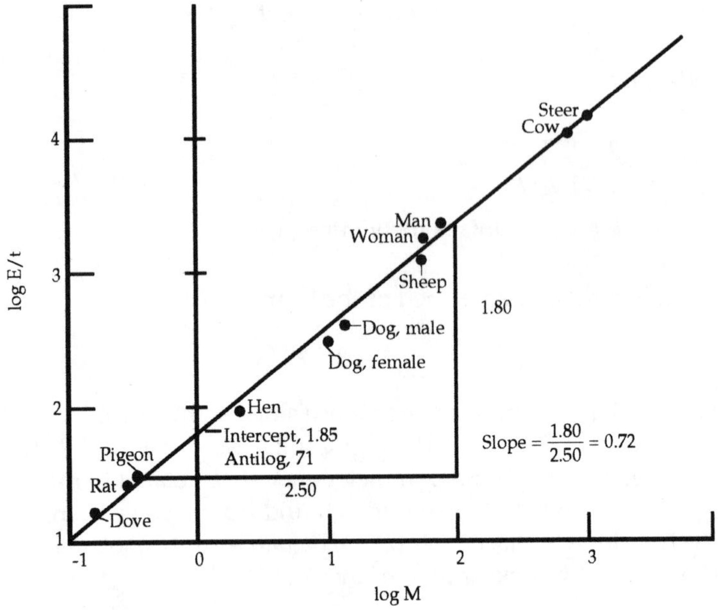

Figure 4.3
The basal metabolic rate (log *E/t)* plotted against *M* for the data of Figure 4.2.

Section 5

LIVING CELLS AS CAPACITORS

Let us consider a feature of living cells that gives them capacitor-like characteristics. As shown in Figure 5.1, the presence of charged ions in a cell and in the fluid surrounding the cell sets up a charge distribution across the membrane wall. From this charge distribution, we see that the cell is equivalent to a small capacitor separated by a dielectric, with the membrane wall acting as the dielectric. The potential difference across this "capacitor" can be measured using the technique described in Figure 5.2. A tiny probe is forced through the cell wall, and a second probe is placed in the extracellular fluid. With this technique, typical potential differences across the cell wall are found to be of the order of 100 mV.

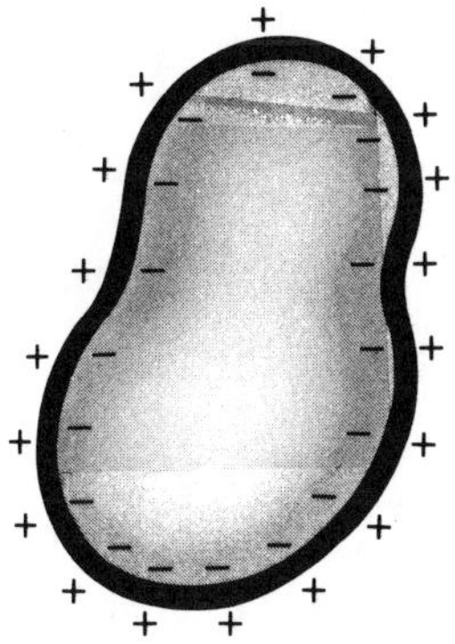

Figure 5.1
A living cell is equivalent to a small capacitor separated by a dielectric, which is the membrane wall of the cell.

The charge distribution of a cell can be understood on the basis of the theory of transport through selectively permeable membranes, discussed in Section 2. The primary ionic constituents of a cell are potassium ions (K^+) and chloride ions (Cl^-). The cell wall is highly permeable to K^+ ions but only moderately permeable to Cl^- ions. Since the K^+ ions can cross the cell wall with ease, the charge distribution on a cell wall is determined primarily by the movement of these ions.

In order to explain the process, let us assume that the cellular fluid is electrically neutral at some point. Under normal circumstances, the concentration of K^+ ions outside the cell is much smaller than inside the cell. Therefore, K^+ ions will diffuse out of the cell, leaving behind a net negative charge. As diffusion commences, electrostatic forces of attraction across the membrane wall produce a layer of positive charge on the exterior surface of the wall and a layer of negative charge on the interior of the wall. At the same time, the electric field set up by these charges *impedes* the flow of additional K^+ ions diffusing out of the cell. Because of this impeding effect to diffusion, an equilibrium is finally established such that there is no net movement of K^+ ions through the cell wall. When this equilibrium situation is reached, the potential difference across the cell wall is given by the **Nernst potential,**

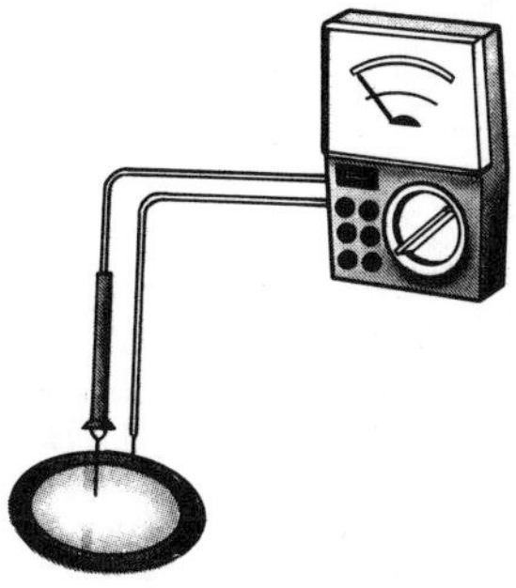

Figure 5.2
An experimental technique for measuring the potential difference across the walls of a living cell.

$$V_N = \frac{kT}{q}\ln\left(\frac{c_i}{c_o}\right) \tag{5.1}$$

where k is Boltzmann's constant (1.38×10^{-23} J/K), T is the Kelvin temperature, q is 1.6×10^{-19} C, and c_i and c_o are the concentrations of K^+ ions inside and outside the cell. At normal body temperatures (T=310 K), this equation becomes

$$V_N = (26.7\text{ mV})\ln\left(\frac{c_i}{c_o}\right) \tag{5.2}$$

The electrical balance achieved in the cell is necessary in order for it to function properly. A threat to this balance occurs when the extracellular fluid contains a higher concentration of sodium ions (Na^+) than is normally found inside a cell.

Although the cell wall is not highly permeable to these ions, a fraction are able to penetrate the wall. This encroachment by Na^+ ions inside the cell will gradually deplete the negative charge on the interior of the cell wall. If this occurs, more K^+ ions will escape, thus endangering the cell. Fortunately, the cell is able to prevent this by effectively "pumping" Na^+ ions out of the cell while "pumping" K^+ ions into the cell. The mechanism for these processes is not fully understood.

Example 5.1 Concentration Ratio

Find the ratio of the concentration of K^+ ions inside a cell to the concentration outside the cell if the Nernst potential is measured to be 90 mV.

Solution

Solving Equation 5.2 for $\ln(c_i/c_o)$, we obtain

$$\ln\left(\frac{c_i}{c_o}\right) = \frac{V_N}{26.7\times10^{-3}} = \frac{90\times10^{-3}\,\text{V}}{26.7\times10^{-3}\,\text{V}} = 3.37$$

The number whose natural logarithm is 3.37 is 29.1. (Use your calculator to verify this.) Thus, the required concentration ratio is

$$\frac{c_i}{c_o} = 29.1$$

Section 6

VOLTAGE MEASUREMENTS IN MEDICINE

6.1 Electrocardiograms

Every action involving the body's muscles is initiated by electrical activity. The voltages produced by muscular action in the heart are particularly important to physicians. Voltage pulses cause the heart to beat, and the waves of electrical excitation associated with the heart beat are conducted through the body via the body fluids. These voltage pulses are large enough to be detected by suitable monitoring equipment attached to the skin. Standard electric devices can be used to record these voltage pulses because the amplitude of a typical pulse associated with heart activity is of the order of 1 mV. These voltage pulses are recorded on an instrument called an **electrocardiograph**, and the pattern recorded by this instrument is called an **electrocardiogram (EKG)**. In order to understand the information contained in an EKG pattern, it is useful to first describe the underlying principles concerning electrical activity in the heart.

The right atrium of the heart contains a specialized set of muscle fibers called the SA (sinoatrial) node, which initiate the heartbeat (Figure 6.1). Electric impulses that originate in these fibers gradually spread from cell to cell throughout the right and left atrial muscles, causing them to contract. The pulse that passes through the muscle cells is often called a *depolarization wave* because of its effect on individual cells. If an individual muscle cell were examined, an electric charge distribution would be found on its surface, as shown in Figure 6.2. (See Living Cells as Capacitors in Section 5 for an explanation of how this charge distribution arises.) The impulse generated by the SA node momentarily changes the cell's charge distribution to that shown in Figure 6.2. The positively charged ions on the surface

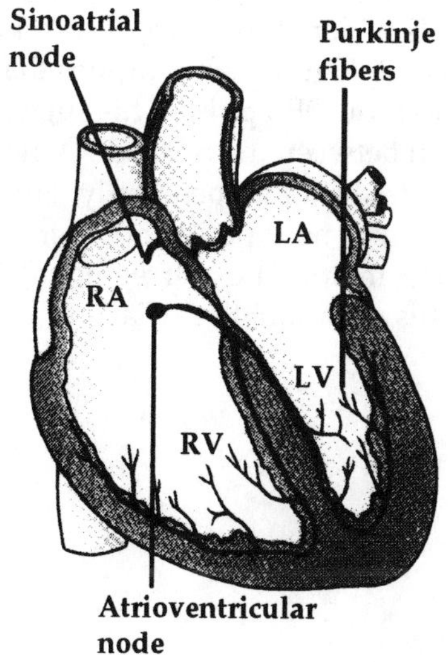

Figure 6.1
The electrical conduction system of the human heart. (RA: right atrium; LA: left atrium; RV: right ventricle; LV: left ventricle.)

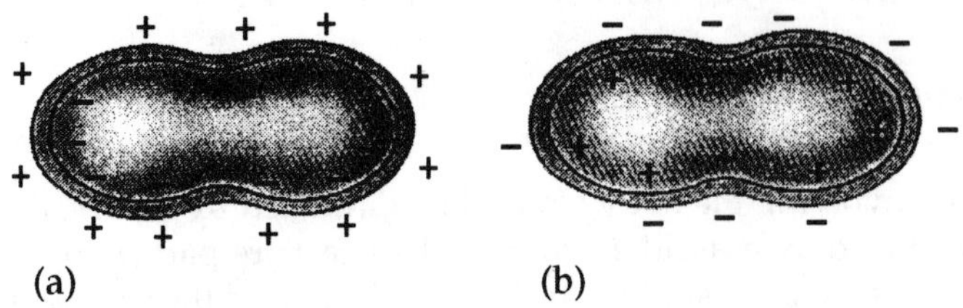

Figure 6.2
(a) Charged distribution of a muscle cell in the atrium before the depolarization wave has passed through the cell. (b) Charge distribution as the wave passes.

of the cell are temporarily able to diffuse through the membrane wall such that the cell attains an excess positive charge on its inside surface. As the depolarization wave travels from cell to cell throughout the atria, the cells recover to the charge distribution shown in Figure 6.2a. When the impulse reaches the AV (atrioventricular) node (Figure 6.1), the muscles of the atria begin to relax, and the pulse is directed by the AV node to the ventricular muscles. The muscles of the ventricles then contract as the depolarization wave spreads through the ventricles along a group of fibers called the *Purkinje fibers*. The ventricles then relax after the pulse has passed through. At this point, the SA node is again triggered and the cycle is repeated.

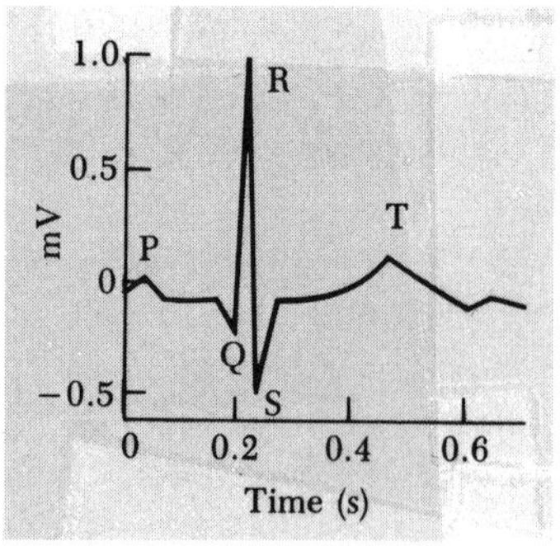

Figure 6.3
An EKG response for a normal heart.

A sketch of the electrical activity registered on an EKG for one beat of a normal heart is shown in Figure 6.3. The pulse indicated by P occurs just before the atria begin to contract. The QRS pulse occurs in the ventricles just before they contract, and the T pulse occurs when the cells in the ventricles begin to recover. EKGs for an abnormal heart are shown in Figure 6.4. The QRS portion of the pattern shown in Figure 6.4a is wider than normal. This indicates that the patient may have an enlarged heart. Figure 6.4b indicates that there is no relationship between the P pulse and the QRS pulse. This suggests a blockage in the electrical conduction path between the SA and AV nodes. This can occur when the atria and ventricles beat independently. Finally, Figure 6.4c shows a situation in which there is no P pulse and an irregular spacing between the QRS pulses. This is symptomatic of irregular atrial contraction, which is called fibrillation. In this situation, the atrial and ventricular contractions are irregular.

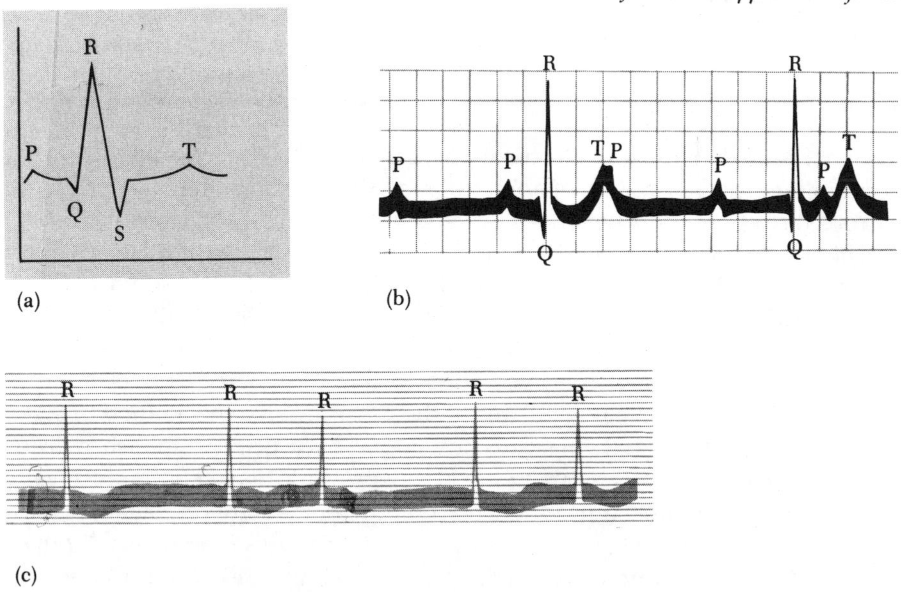

Figure 6.4
Abnormal EKGs.

6.2 Electroencephalography

The electrical activity of the brain can be measured with an instrument called an **electroencephalograph** in much the same way that an electrocardiograph measures the electrical activity of the heart. The voltage pattern measured by an electroencephalograph is referred to as an **electroencephalogram (EEG)**. An EEG pattern is recorded by placing electrodes on the patient's scalp. While an EKG voltage pulse is typically 1 mV, a typical voltage amplitude associated with brain activity is only a few *micro*volts and is therefore more difficult to measure.

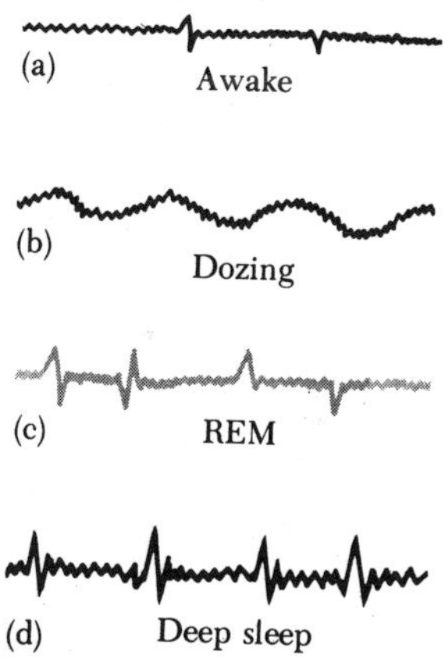

Figure 6.5
Brain waves from an individual in various stages of sleep.

The EEGs in Figure 6.5 represent the brain wave pattern of a patient awake and then in various stages of sleep. In Figure 6.5b the patient begins to fall into a light sleep, and in Figure 6.5d the patient is in a deep sleep. Figure 6.5c is the EEG during a type of sleep called **REM (rapid eye movement)** sleep, which occurs approximately every 2 h. In this stage, the brain activity is quite similar to that of the patient while awake. It is interesting to note that a person in this stage of sleep is extremely difficult to awaken. Apparently, **REM** sleep is necessary for psychological well-being. Anyone who is deprived of this stage of sleep for an extended period of time becomes extremely fatigued and irritable.

The EEG is an extremely important diagnostic tool for detecting epilepsy, brain tumors, brain hemorrhage, meningitis, and so forth. For example, the EEG pattern of a patient suffering an epileptic seizure would show very little structure, indicating that the brain activity is greatly reduced.

Brain waves are often discussed in terms of their frequencies. Frequencies of about 10 Hz are referred to as *alpha waves*, frequencies between 10 Hz and 60 Hz are called *beta waves*, and those below 10 Hz are called *delta waves*. As a person falls asleep, the frequency of brain wave activity generally decreases. For example, the brain wave frequency for a person in a very deep sleep may drop as low as 1 or 2 Hz.

ESSAY

Current in the Nervous System

Paul Davidovits
Boston College

The most remarkable use of electrical phenomena in living organisms is found in the nervous system of animals. Specialized cells in the body called neurons form a complex network that receives, processes, and transmits information from one part of the body to another. The center of this network is located in the brain, which has the ability to store and analyze information. Based on this information, the nervous system controls parts of the body.

The nervous system is very complex: the human nervous system, for example, consists of about 10^{10} interconnected neurons. Some aspects of the nervous system are well known. During the past 35 years, the method of signal propagation through the nervous system has been firmly established. The messages are electric pulses transmitted by neurons. When a neuron receives an appropriate stimulus, it produces electric pulses that are propagated along its cable-like structure. The strength of the stimulus is conveyed by the number of pulses produced. When the pulses reach the end of the "cable," they activate other neurons or muscle cells.

The neurons, which are the basic units of the nervous system, can be divided into three classes: sensory neurons, motoneurons, and interneurons. The sensory neurons receive stimuli from sensory organs that monitor the external and internal environment of the body. Depending on their specialized functions, the sensory neurons convey messages about factors such as heat, light, pressure, muscle tension, and odor to higher centers in the nervous system. The motoneurons carry messages that control the muscle cells. The messages are based on the information provided by the sensory neurons and by the brain. The interneurons transmit information from one neuron to another.

Each neuron consists of a cell body to which are attached input ends called **dendrites** and a long tail called the **axon**, which propagates the signal away from the cell (Fig.1). The far end of the axon branches into nerve

Figure 1
Diagram of a neuron.

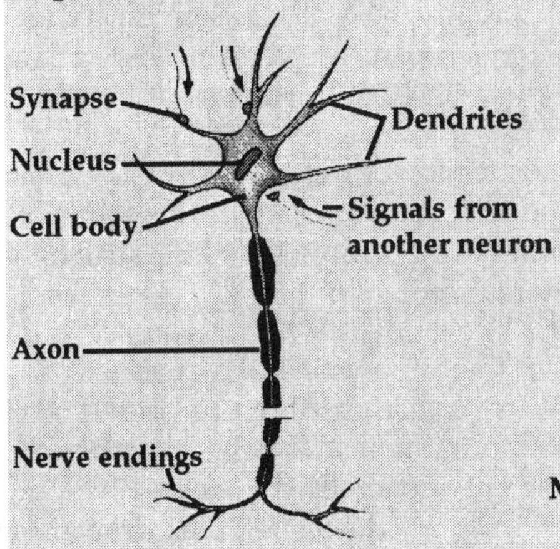

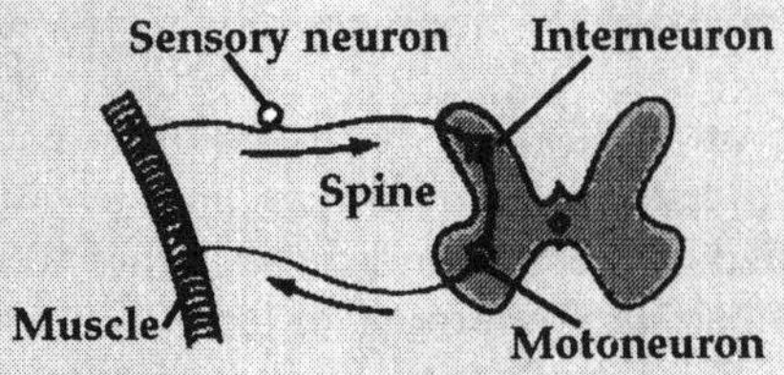

Figure 2
A simple neural circuit.

endings that transmit the signal across small gaps to other neurons or to muscle cells. A simple sensory-motoneuron circuit is shown in Figure 2. A stimulus from a muscle produces nerve impulses that travel to the spine. Here the signal is transmitted to a motoneuron, which in turn sends impulses to control the muscle.

The axon, which is an extension of the neuron cell, conducts the electric impulses away from the cell body. Some axons are extremely long. In humans, for example, the axons connecting the spine with the fingers and toes are more than 1 m long. The neuron can transmit messages because of the special electrical characteristics of the axon. Most of the information about the electrical and chemical properties of the axon is obtained by inserting small needle-like probes into the axon. With such probes it is possible to measure currents flowing in the axon and to sample its chemical composition. Such experiments are usually difficult to run because the diameter of most axons is very small. Even the largest axons in the human nervous system have a diameter of only about 20×10^{-4} cm. The giant squid, however, has an axon with a diameter of about 0.5 mm, which is large enough for the convenient insertion of probes. Much of the information about signal transmission in the nervous system has come from experiments with the squid axon.

In the aqueous environment of the body, salts and other molecules dissociate into positive and negative ions. As a result, body fluids are relatively good conductors of electricity. The inside of the axon is filled with an ionic fluid that is separated from the surrounding body fluid by a thin membrane that is only about 5 nm to 10 nm thick.

The resistivities of the internal and external fluids are about the same, but their chemical compositions are substantially different. The external fluid is similar to seawater. Its ionic solutes are mostly positive sodium ions and negative chloride ions. Inside the axon, the positive ions are mostly potassium ions and the negative ions are mostly large negatively charged organic ions.

Since there is a large concentration of sodium ions outside the axon and a large concentration of potassium ions inside, we may ask why the concentrations are not equalized by diffusion. In other words, why don't the sodium ions leak into the axon and the potassium ions leak out of it? The answer lies in the properties of the axon membrane.

In the resting condition, when the axon is not conducting an electric pulse, the axon membrane is highly permeable to potassium ions, slightly permeable to sodium ions, and impermeable to the large organic ions. Thus, while sodium ions cannot easily leak into the axon, potassium ions can certainly leak out of it. As the potassium ions leak out of the axon, however, they leave behind large negative ions, which cannot follow them through the membrane. As a result, a negative potential is produced inside the axon with respect to the outside. The negative potential, which has been measured to about 70 mV, holds back the outflow of potassium ions so that, at equilibrium, the concentration of ions is as we have stated.

The mechanism for the production of an electric signal by the neuron is conceptually remarkably simple. When a neuron receives an appropriate stimulus, which may be heat, pressure or a signal from another neuron, the properties of its membrane change. As a result, sodium ions first rush into the cell while potassium ions flow out of it. This flow of charged particle constitutes an electric current signal which propagates along the axon to its destination.

Although the axon is a highly complex structure, its main electrical properties can be represented by the standard electric circuit concepts of resistance and capacitance. The propagation of the signal along the axon is then well described by the techniques of electric circuit analysis discussed in the text.

Suggested Reading

B. Katz, "How Cells Communicate," *Sci. American*, September 1981m, p. 208.
B. Katz, *Nerve Muscle and Synapse*, New York, McGraw-Hill Inc., 1966.

Essay Questions

1. Why has the nervous system developed to utilize sodium and potassium ions to conduct electrical signals?

2. In the nervous system the strength of a stimulus is conveyed by the number of pulses produced rather than by the amplitude of the signal. What is the advantage of this arrangement?

Section 7

COLOR

The wavelengths in the visible portion of the electromagnetic spectrum range from 400 nm to 700 nm, with the shortest wavelength corresponding to violet and the longest corresponding to red. Each wavelength between these two extremes has its own characteristic hue, or shade. The spectrum of colors can be remembered by the boy's name "ROY G BIV," where each letter represents one of the prominent spectral colors: red, orange, yellow, green blue, indigo, and violet. (Indigo is not really distinct from violet, but it does allow the boy's last name to have a vowel.) If all the colors of the visible spectrum are incident on the eye simultaneously, the mixture is interpreted as white by the eye. On the other hand, black denotes the complete absence of color.

A thorough study of color and of why we see colors as we do would constitute a rather long chapter or perhaps even a textbook to do it justice. In this section, we shall present some concepts that will help you have some understanding of the colors in our world.

The color of an object depends on one or more of the following three processes:

1. The emission of light by the object
2. The reflection of light by the object
3. The transmission of light by the object

7.1 Emission of Light

Any object whose atoms have been sufficiently excited by the absorption of energy will emit light. For example, if you heat an iron poker to the point where it glows, the added heat energy causes a rearrangement of the electrons in the atoms of the poker, and light is emitted by the atoms as they return to the state they were in before the rearrangement took place. Hence, the light we see when the poker glows arises from this emission.

Likewise, the light from a neon sign in a store window is emitted by the excited gas atoms contained in the tubes. The colors emitted by a heated gas depend on the characteristics of atoms contained in the tubes.

7.2 Transmission of Light

The color of a piece of transparent material depends, to a large extent, on the wavelengths of light that the material transmits. A piece of colored glass receives its color by absorbing most of the wavelengths of light incident upon it while allowing others to pass through. Those wavelengths that pass through (or are transmitted) give the glass a distinctive color. For example, a piece of red glass in the stained glass window of a church appears red because it absorbs all colors in the visible spectrum except those the eye interprets as red. Because the various wavelengths of light carry energy, a piece of glass is slightly warmed when it absorbs light. A clear piece of glass, such as a window pane, allows all colors in the visible spectrum to pass through and so has no characteristic color.

Pieces of colored glass are often used as optical filters to select a particular wavelength for certain experiments. A color camera uses three filters to separate the light from a scene into the three primary colors – red, blue, and green.

7.3 Reflection of Light

Most of the objects around us achieve their color by reflection. For example, a red rose neither emits nor transmits visible light. Rather, its color arises because it reflects only those wavelengths in the visible spectrum that are interpreted by the eye as red. Thus, when a red rose is placed in sunlight, it absorbs all the colors in the visible spectrum except red. Now imagine that we take our red rose into a room that is illuminated with only a green light source. There is no red light available from the source, and hence the rose will not reflect any light at all. For this reason, the rose will appear to be black when viewed with green light. On the other hand, the stem and leaves of the rose will appear to have their natural green color when viewed in green light. Before you read the next few sentences, see if you can guess what the American flag would look like if you observed it in a room illuminated with only a blue light source. The red stripes would appear black because there is no red light for them to reflect. The white stripes and stars would appear blue because all the colors in the spectrum are reflected by a white object. Finally, the blue square in the corner would appear blue because there is blue light in the room for it to reflect.

7.4 The Additive Primaries and Color Vision

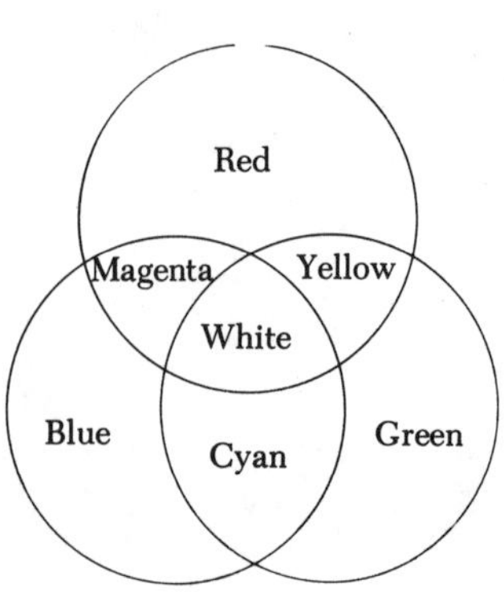

Figure 7.1
The color wheel.

When all the colors of the visible spectrum are mixed together, the light appears to be white. However, if equal intensities of red, green and blue light are mixed together, the result is also interpreted by the eye as white. In fact, *all of the colors in the visible spectrum can be generated by mixing together proper proportions of these three colors.* Because of their ability to reproduce the sensation of any color when added to each other in varying amounts, red, green and blue are called the additive primaries.

You can demonstrate some of the properties of the additive primaries by placing a piece of red cellophane over the face of one flashlight, a piece of green cellophane over the face of a second flashlight, and a piece of blue cellophane over the face of a third flashlight. When the three beams are superimposed on a white wall, the illuminated area will appear as in Figure 7.1.

The area where the green light overlaps the red light appears yellow, a magenta area appears where the red light overlaps the blue, and a cyan (bluish green) area appears where the green light overlaps the blue. However, the central region, where all the three colors overlap, is white.

The color television set is a good example of how to make practical use of color mixing. The colors observed on the screen of a color television set are produced when the electron beams strike the screen causing the dots to fluoresce in various colors. If three dots at a particular location on the screen are emitting equal intensities of light, the blue, red and green light coming from this location appear to be white light. If the blue dot at another location is emitting no light while the red and green dots are emitting light of equal intensity, that spot appears to be yellow.

The light sensors on the retina of the eye are of two types, rods and cones. The rods are sensitive enough to respond to dim light and to small variations in light intensity. However, they cannot distinguish between the various wavelengths, or colors, of light. The cones are the sensors that enable us to distinguish colors. There are three types of cones, each sensitive to one of the three additive primary colors. The cones sensitive to red light respond when illuminated by red, orange, yellow, and, to a lesser extent, green light. The signal sent to the brain from the red-sensitive cones is the same regardless of which of these colors strikes the cones. Likewise, the green-sensitive cones respond primarily to green and yellow light and, to a lesser extent, blue and orange light. Finally, the blue-sensitive cones are sensitive to blue and violet light.

Now let us assume that light entering the eye stimulates only the red-sensitive cones, which in turn causes a message to be sent to the brain. Based on this information alone, the brain would have no way of determining the color of the light. In addition, however, the brain recognizes that no signals are being sent by the green-sensitive cones. Hence, the brain concludes that the light must be red since it receives signals from the red-sensitive cones only. If pure yellow light enters the eye, it stimulates both the red-sensitive and the green-sensitive cones. The brain interprets such stimulation as yellow. It is interesting that light of this single wavelength appears to be yellow, even though we see the same yellow color when a combination of green and red light strikes the retina. Such a combination appears to be yellow even though there is no yellow light present. This is explained by the fact that red and green wavelengths stimulate both the red-sensitive and green-sensitive cones in the same manner as light of a single yellow wavelength.

Most animals are completely color blind. Except for the other primates and a few other species, notably bees, the human being is the only animal capable of seeing the world in full color. Complete lack of color vision is very rare in humans, but partial color blindness is common. About one man in 12 and one woman in 120 have what is known as red-green blindness. The eyes of such individuals respond in the same way to red and green light, and therefore the brain cannot distinguish these colors.

Section 8

THE PHYSICS OF VISION

8.1 Chromatic Aberration and the Eye

The dispersion of light by glass results in a phenomenon in lenses called *chromatic aberration*. When chromatic aberration is present, images show colored edges. The phenomenon is illustrated in Figure 8.1a, in which the light is shown being allowed to pass through only the edge of a lens; this is not unlike a prism. The red is deviated the least, the blue the most. In Figure 8.1b the ideal image of an object that is half black and half white is shown. If there is an obstacle to allow the light in through only the edge of the lens, the dispersion will cause the boundary to appear as a blue line protruding into the white of the image as in Figure 8.1c. If an object were inverted, as in Figure 8.1d, the edge would have a red tinge protruding into the white. A color corrected or achromatic lens will not show such colored edges because it consists of two lenses of different dispersive powers. The first lens separates the colors, and the second one focuses the separate colors together at the focal distance as in Figure 8.1e. If the second lens over-corrected the dispersion of the first, the situation would be as shown in Figure 8.1f; with the white at the bottom of the object as shown, the blue would appear in the white part of the edge.

The phenomenon of dispersion and the extent of the correction for chromatic aberration in the human eye can be simply demonstrated. While you are viewing a field with a dark and a bright area as in the various diagrams of Figure 8.1 or as in Figure 8.2, raise a small card just in front of the eye to allow the light through just the top of the cornea and lens. With the dark area on top, as could be obtained by viewing a white bulb or light fixture, the nature of the image on the retina that we see (except for the inversion by the brain) is illustrated in Figure 8.2. If there is no correction for dispersion, the red would appear in the white as in Figure 8.1d. If there is optimum correction, there would be no color along the edge as in Figure 8.1e; and if the eye over-corrects, then the blue should appear in the white as in Figure 8.1f. The latter situation has been included only because when this experiment is performed the blue edge appears in the white, indicating an *over-correction* for dispersion. This is a simple experiment that you should do to see the effect with your own eye. It can be repeated with the object oriented as in Figure 8.1c, moving the card upward in front of the eye and finding whether the blue or the red is into the white.

Figure 8.1
Effects of dispersion in lenses. In these examples the light is allowed to go through only the edge of a lens. The phenomenon is shown in (a), and (b) shows the formation of an ideal image by a lens. Two examples of effects of dispersion are shown in (c) and (d), and (e) illustrates the elimination of the effect with an achromatic lens. Overcorrection for dispersion is shown in (f), which can be compared to the uncorrected lens in (d).

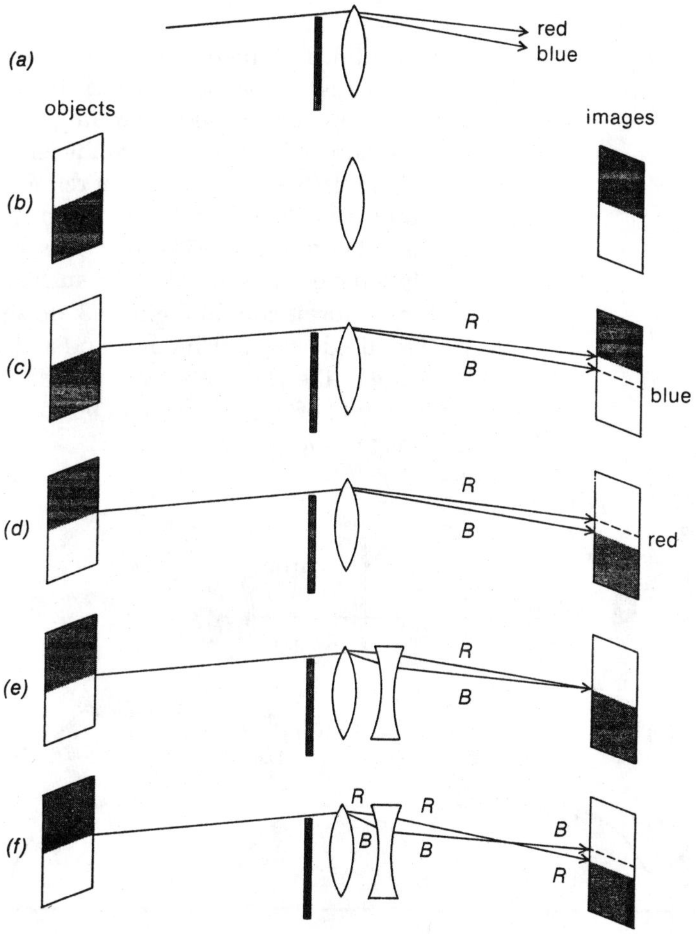

Figure 8.2
Chromatic aberration in the eye. Compare this to Figure 8.1 (d), (e), and (f).

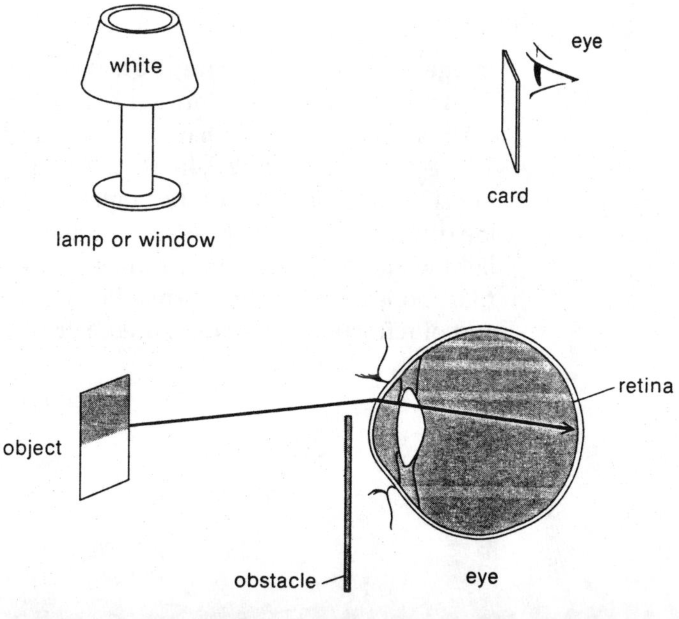

8.2 The Eye

The human or mammalian eye is often compared with a camera, and in many ways this comparison is valid. There are some differences, however, which are important for understanding the factors affecting proper image formation. The camera has air between the lens and the film, whereas the eye has material of relatively high index of refraction filling the whole eyeball from the front surface to the retina. In animals that live in air, most of the refraction used to produce the image occurs at the front surface, the cornea. The eye does have a lens also, but for such animals it is of secondary importance for image formation. In Figure 8.3 are shown the camera, the eye, a simplified or "reduced" eye, a fish bowl, and a glass sphere. The eye resembles the fish bowl or the glass sphere more than it does the camera, so the image-forming property of one surface will be considered first, and then the formation of an image by a sphere or ball.

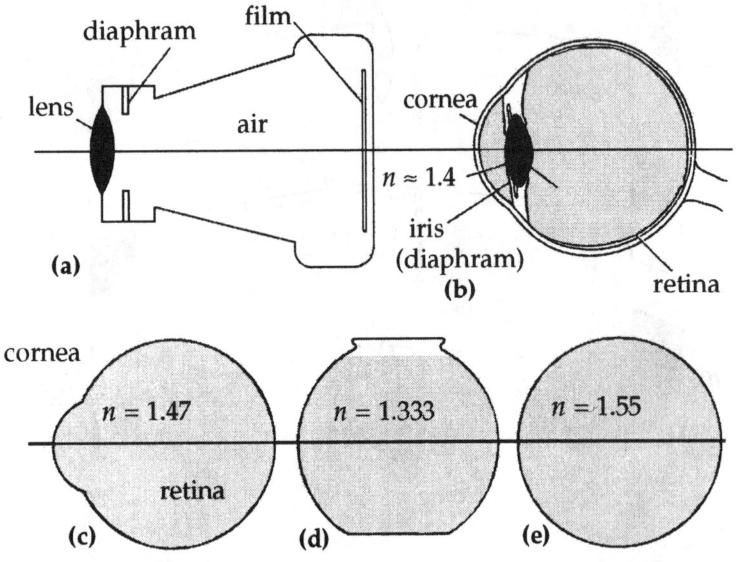

Figure 8.3
Diagrams to compare and contrast **(a)** a camera, **(b)** an eye, **(c)** a reduced eye, **(d)** a fish bowl, and **(e)** a glass sphere.

Image Formation by a Single Surface

A curved surface which forms a boundary between two media of different indices of refraction can have a focusing property similar to that of a lens. This is shown in Figure 8.4a. Figure 8.4b is a similar diagram, but only one ray is shown and some of the angles are labeled. The radius of curvature in the direction shown is R. A surface that was concave toward the incident light would have a negative radius. The radius line meets the surface normal to it and is therefore the normal line from which the angles of incidence (i) and of refraction (r) are measured. For two media, Snell's law is used in the form

$$n_1 \sin i = n \sin r \tag{8.1}$$

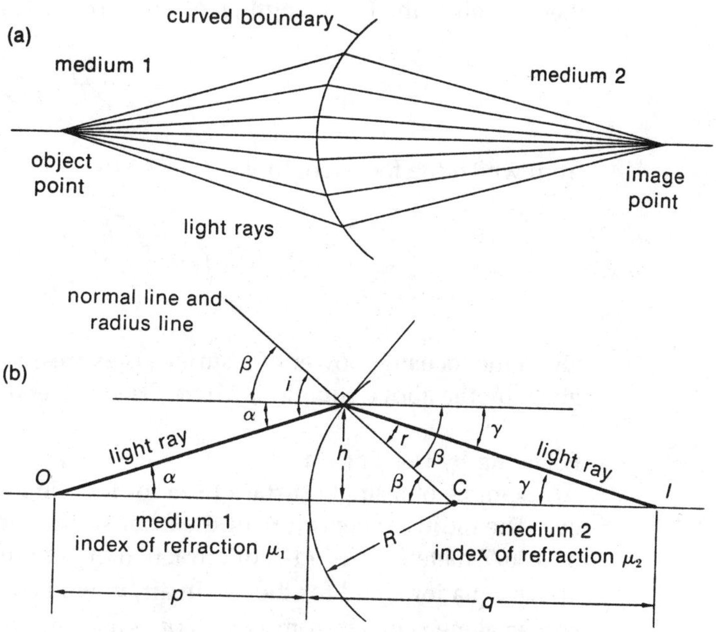

Figure 8.4
Focusing by a single curved surface. (a) Overall view. (b) Single ray.

The angles α, β, and γ as shown along the axis are used in the analysis. On examining the figure, it is seen that $i=(\alpha + \beta)$ and $r=(\beta - \gamma)$. If we assume that all the angles are small, then the sine of the angle is equal to the angle (in radians); so the above form of Snell's law adapted to this situation reduces to

$$n_1 (\alpha + \beta) = n_2 (\beta - \gamma)$$

The angles α, β, and γ in radian measure (the angles must again be recognized to be small) are given by h/p, h/R, and h/q respectively. The last equation then becomes, with the h cancelled,

$$n_1\left[\frac{1}{p} + \frac{1}{R}\right] = n_2\left[\frac{1}{R} - \frac{1}{q}\right]$$

This can be manipulated into the form

$$\frac{n_1}{p} + \frac{n_2}{q} = \frac{n_2 - n_1}{R} \qquad \textbf{(8.2)}$$

The fact that h, the distance of the ray from the axis, canceled out means that all the rays emanating from the point at the distance p and hitting the surface at any distance h from the axis are focused to the same point at the distance q. Thus, an object point at O is focused at the position I

The equation is beautifully general, and it will be used to analyze a few physical situations. First, let the object be a long distance away so that n_1/p approaches zero. The rays approaching the surface will be parallel, and the distance q can then be called the focal length, f, of the surface. The equation then becomes

$$\frac{n_2}{f} = \frac{n_2 - n_1}{R}$$

from which the focal length is

$$f = \frac{n_2 R}{(n_2 - n_1)} \qquad \textbf{(8.3)}$$

Often the focusing power of a surface is expressed in diopters (that is, $1/f$ as given by the above equation, where f is in meters).

Focusing by the Cornea

An example of a single surface used for focusing is the cornea of the human eye. The radius of curvature of the front surface of the cornea is about 7.7 mm or 0.0077 meters. The index of refraction of the human eye is 1.336 (n_2 in the above equation); and for the eye in air, n_1 is 1.000. The focal length of the cornea alone is given from $f=n_2R/(n_2 - n_1)$, where n_2 = 1.336, n_1 = 1.000 (air), and R = 7.7 mm. Putting these numbers into the equation yields a value of f = 30 mm. This is the distance from the cornea to the position at which a distant object would be focused. The normal length of the eyeball is about 24 mm. If there were no lens, the cornea alone would focus an image only 6 mm or a quarter of an inch behind the retina, as illustrated in Figure 8.5a; the image position would vary with the object distance. The lens can normally adjust its curvature to provide sharp focusing for objects as far as infinity and as close as 10 inches. If the lens is removed, vision will be blurred; but spectacles can be used to provide the extra focusing normally done by the lens of the eye, as shown in Figure 8.5c. The power of the spectacles in such a case must be quite high, amounting to a focal length of about 4 inches.

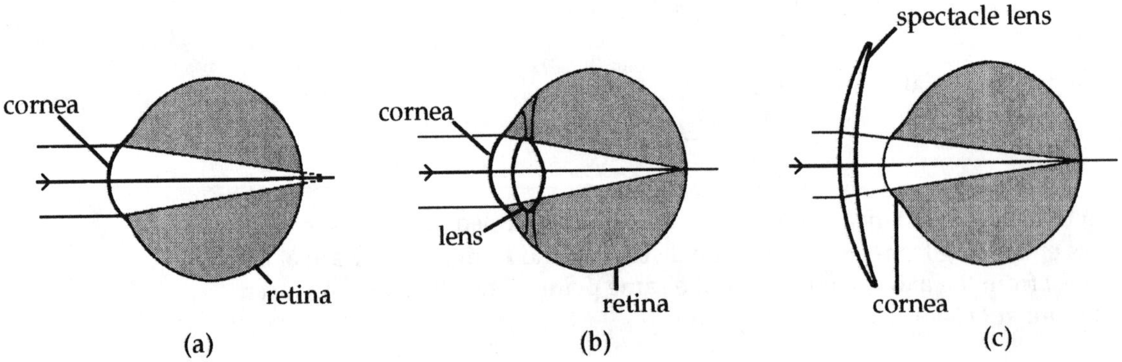

Figure 8.5
(a) Focusing by cornea alone. (b) Focusing by cornea and lens. (c) Focusing by spectacles plus cornea.

The Reduced Eye

There are several refracting surfaces in the human eye, and the calculation of image sizes and properties by considering each of them is a long complex process. Most of the focusing occurs at the corneal surface, and a simplified model of the eye can be constructed on the basis of the assumption that *all* the focusing occurs at the corneal surface. If the material of the eyeball had an index of 1.47 rather than about 1.336, this would be the case. What is referred to as the **reduced eye** maintains the same dimensions as the real eye; but for image calculations it is considered to be filled with a uniform material of an index that will result in the image being formed on the back surface (retina). The length of the eyeball averages about 24 mm, and the corneal radius is about 7.7 mm. The retinal image properties calculated for the reduced eye are very close to those for the actual eye.

The reduced eye is illustrated in Figure 8.3c and in Figure 8.6. In Figure 8.6a, some of the image-forming light rays are also shown. Rays of interest are those which go through the center of curvature of the cornea. These rays are normal to the corneal surface and are therefore not deviated. The relation between image size and object size is found from the shaded triangle of Figure 8.6b. This is

$$\frac{I}{O} = \frac{16.3\text{ mm}}{p + 7.7\text{ mm}}$$

where p is measured from the corneal surface. The 7.7 mm can be neglected; and since eyeballs do vary in size, one obtains that, to a very close approximation,

$$\frac{I}{O} = \frac{16\text{ mm}}{p} \qquad \textbf{(8.4)}$$

This relation will be used in the study of visual acuity, in which the image size and receptor cell size are considered.

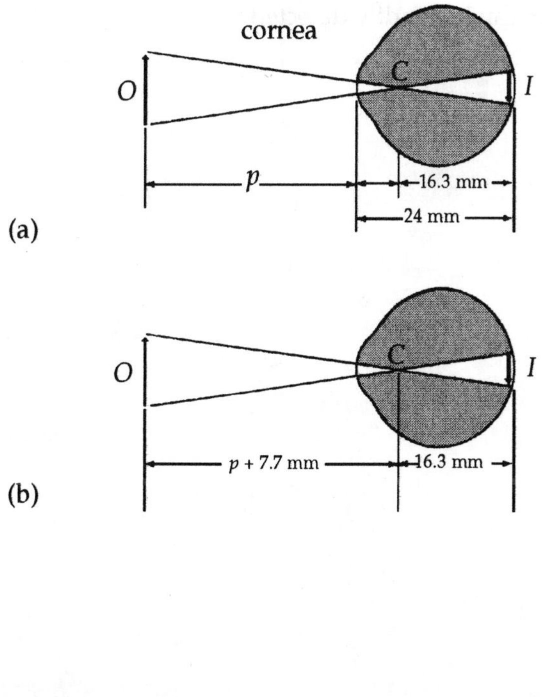

Figure 8.6
Diagrams to show how to find the image size on the retina using the reduced eye.

EXAMPLE 8.1

If a person 1.7 meters high (5′7″) is viewed from a distance of 6 meters, what would be the size of the image on a retina?

Use $\frac{I}{O} = \frac{16\text{ mm}}{p}$ or $I = \frac{16\text{ mm}}{p} \cdot O$

with $p = 6$ m and $O = 1.7$ m.

The angular size will not be small, and a small error will be recognized to occur because of this. Solving,

$$I = \frac{16\text{ mm} \times 1.7\text{ m}}{6\text{ m}} = 4.5\text{ mm}$$

The image on the retina is 4.5 mm high.

EXAMPLE 8.2

If two small objects 1 mm apart are viewed from 25 cm or 250 mm, what would be the separation of the images? The same relations as in Example 8.1 are used:

$$I = \frac{16\text{ mm} \times 1\text{ mm}}{250\text{ mm}}$$

$$= 0.064\text{ mm or }64\ \mu\text{m}$$

This example shows that the images on the retina are very small and that detail measured in μm can be readily detected.

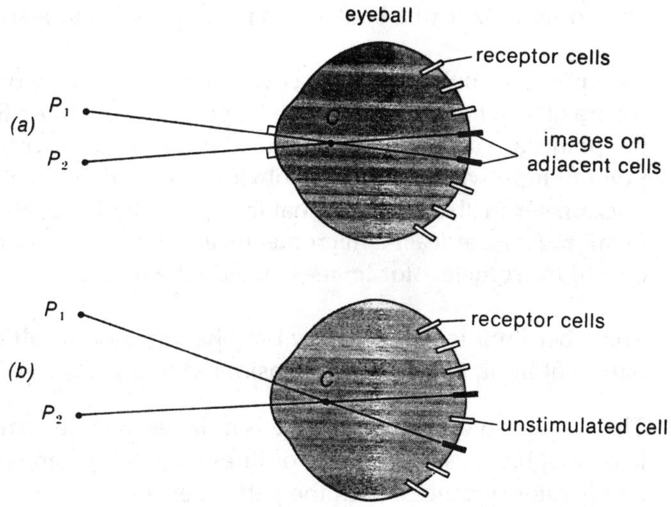

Figure 8.7
Resolution by the eye. In (a) the images are on adjacent receptor cells and are seen as one. In (b) there is an unstimulated cell between the images, so they are seen as two.

8.3 Sharpness of Vision

A newspaper picture consists of a series of dots of different sizes, very close together, which as a whole give the various tones to the picture. On careful examination with the unaided eye, the dots can just be seen. Higher quality magazines use much finer dots to make up the picture; without the aid of a lens used as a simple microscope; the dots cannot be seen. It is easy to say that they are just too close together to see, but it is profitable to question why this limit of sharpness of vision exists. One of the reasons for the limitations on the detail that can be seen with the eye is discussed in this section.

There are two basic limiting factors to visual acuity other than imperfect focusing. One is that the retina consists of a large number of closely spaced light-sensitive cells. These cells vary slightly in size and spacing. In the most acute part of the eye (the fovea centralis), where the color-sensitive cells are packed most tightly, they are about a micrometer between centers. Outside this region they are 3 to 5 micrometers apart. Also, away from the fovea are the rod cells used in night vision. They are not color-sensitive, but they respond to extremely low levels of illumination. They also respond in networks; images resulting from the stimulation of rods are not sharp. Outdoors in bright starlight, one may be able to see a page of a book but it is not possible to read it.

Two points of light on an object will not be seen as two distinct points unless the images fall on non-adjacent receptors. As in Figure 8.7, if the images of the points are on adjacent receptor cells, or on the same cell, then only one point of light will be seen. It is obvious, then, that detail which is only a micrometer in size on the retinal image cannot be resolved by the eye; the detail must be at least 2 micrometers across for the most sensitive region and 6 to 10 micrometers for images outside the fovea.

The other limit to visual acuity is a phenomenon resulting from the wave nature of light, and it will be considered in the appropriate section.

The resolution of the normal eye is quite easy to determine experimentally, at least roughly. A series of dots or lines on a paper can be viewed at greater and greater distances, until the pattern eventually loses its detail. Experiments of this sort show the limit of resolution of the eye to be less than one minute of angle (0.0003 radian). This limit for the discrimination of fine detail amounts to just three thousandths of an inch difference at a viewing distance of 10 inches. Under optimum conditions, resolutions down to 30 seconds or even 20 seconds of angle can be obtained. Actual resolution also depends on the brightness of illumination and on the contrast between the various parts of the object.

What is called **visual acuity** is the inverse of the angular resolution in minutes. Since the normal resolution is about 1 minute of angle, the normal acuity is 1. A person who can resolve only 2 minutes of angle has a visual acuity of 1/2: the eyes of that person are only half as "sharp" as normal.

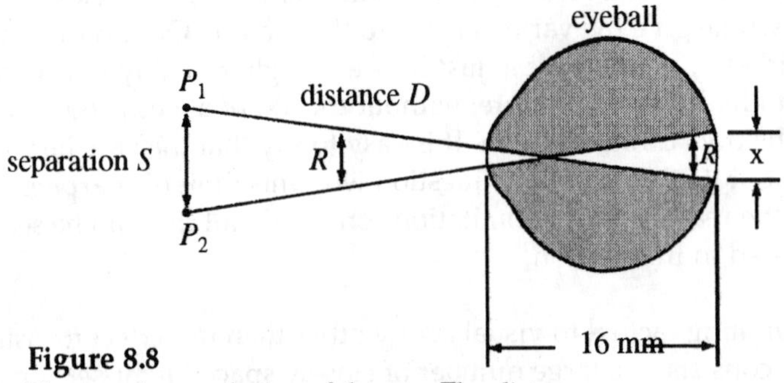

Figure 8.8
The angular resolution of the eye. The distance x is about the spacing of two receptor cells.

Table 8.1 Approximate angular resolution of a human eye for different receptor cell spacings.

Receptor Cell Spacing	Necessary Image Spacing, x	Expected Resolution, R
1 *nm*	2 *nm*	25 seconds
3 *nm*	5 *nm*	1 minute
5 *nm*	10 *nm*	2 minutes

Example 8.3

What are the approximate resolution and the acuity for a person who just ceases to distinguish the millimeter rulings on a meter stick at a distance of 1.6 meters? Also, find the spacing of the 1 mm lines on the retinal image. A 1 mm object spacing at 1.6 meters (1600 mm) is an angle of 1 mm/1600 mm = 0.00062 radian. Since 1 minute = 0.0003 radian, the angular resolution is just over 2 minutes. The visual acuity is 0.5.

If the object size is 1 mm, the image size on the retina is given by

$$I = \frac{16 \text{ mm}}{p} \cdot O$$

where p = 1600 mm. Since O = 1 mm,

$$I = \frac{16 \text{ mm} \times 1 \text{ mm}}{1600 \text{ mm}} = \frac{1}{100} \text{ mm} = 10\ \mu\text{m}$$

The images of the 1 mm lines are spaced at 10 mm on the retina.

You should see for yourself at what distance you can distinguish 1 μm lines on a ruler to *estimate* your own acuity. This is only a crude method. Why?

Resolution Based on Receptor Cell Spacing

In the fovea the receptor cells are about 1 *nm* between centers, and image points must be separated by about 2 *nm* to be seen as distinct points. On other parts of the retina where the receptor cells are farther apart, two image points would have to be separated by 5 or even 10 *nm* to be seen as two. If this separation is called x, as in Figure 8.8, the corresponding angular separation of object points, R, is given in radian measure by

$$R = \frac{x}{16 \text{ mm}} = \frac{x}{16000\ nm}$$

The values of R for different image spacings on the retina are shown in Table 8.1, in which the angle of resolution R has been changed from radians to minutes or

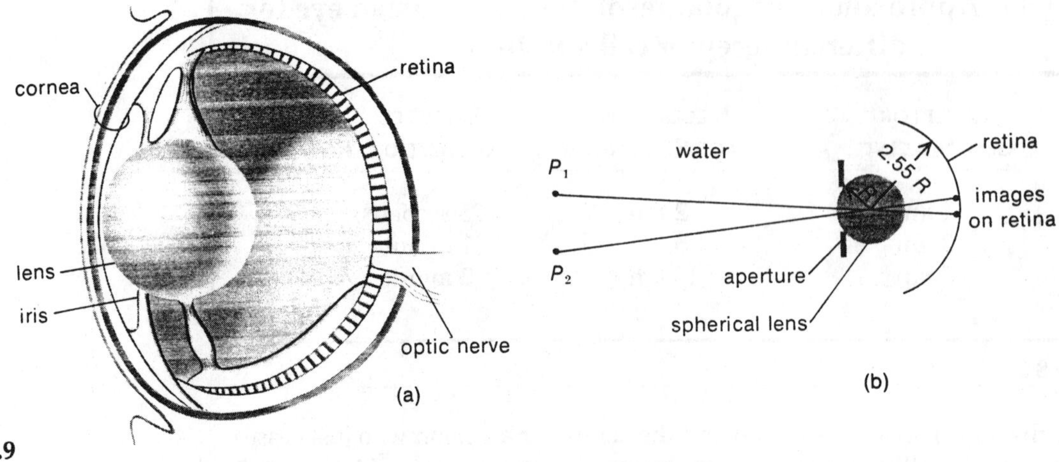

Figure 8.9
The eye of the fish. Note the large spherical lens. Diagram (a) is adapted from G.L. Walls, *The Vertebrate Eye and Its Adaptive Radiation*, Cranbrook Institute of Science, 1942. Diagram (b) is a model of the fish eye which can be used to find properties of the image.

seconds of angle. The angular resolution or acuity based on receptor cell spacing agrees very well with the observed values. Even though these results seem to settle the question of the reason for the limit of resolution by the human eye, the other effect (which is due to the wave nature of light) is worthy of analysis to see if it does play any part in this phenomenon.

8.4 Eyes In Air and Water

Under water the human eye cannot focus sharply on anything. This is because most of the focusing normally takes place at the corneal surface, where there is air of index 1 against the material of the aqueous humor, which is of index 1.336. If the outside material is replaced by water of index 1.333, there is practically no focusing at that surface. The only way that man can see distinctly under water is to use goggles with a flat glass surface. There is then air against the cornea and focusing can take place. The phenomenon of apparent depth must still be taken into account when judging the position of objects seen through the goggles.

The Eye of the Fish

In the eyes of underwater creatures such as fish, there can be practically no focusing at the corneal surface, for the index of refraction of the humors of the eye is almost identical to that of water. The focusing must therefore be done entirely by the lens, which must be of much higher power than the lens of a human eye. This is in fact the case. Figure 8.9 shows the construction of a typical fish eye. The lens is spherical and of high index of refraction. The effective power of the lens is also increased because it has a higher index at

Box 8.1 FOCUSING BY A SPHERE

The focusing properties of a sphere are worthy of a little examination. A glass ball or a spherical flask of water has focusing properties similar to those of the lens of a fish eye, and not unlike even those of a human eye. It is very easy to demonstrate this in the laboratory. If a flask of water is placed at some distance from windows or lights and a piece of white paper is moved close to it, an image will be seen on the paper. If the light entering the flask passes through about a two-inch-diameter hole (a pupil) in a card (iris), the image will be amazingly sharp. The situation is illustrated by ray diagrams in Figure 1, which is based on a sphere of index 1.55 in air. For a sphere of water, $n = 1.33$, the image will be somewhat farther from the sphere than for glass. The distance at which focusing occurs can be expressed in terms of the radius.

To calculate the image position, the equation for focusing by a single surface is used. With the quantities shown in Figure 1, the equation is

$$\frac{n_1}{p} + \frac{n_2}{q} = \frac{n_2 - n_1}{R}$$

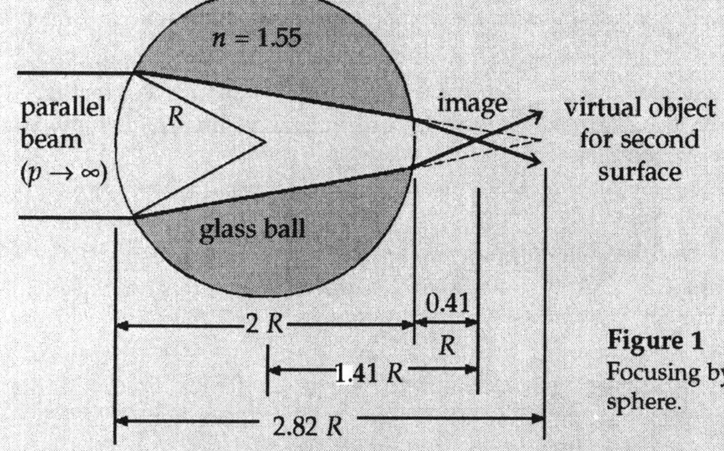

Figure 1
Focusing by a glass sphere.

Consider, for example, a glass sphere in air; for the first surface n_1 is the index for air, just 1.00 and the index n_2 is that of the glass, which typically is about 1.55. The focal length of the first surface as found from the above equation is 2.82R; the rays would be focused beyond the second surface, as is shown in Figure 1. The second surface introduces more bending, so the rays are focused to the point shown as *I*. No matter what the direction of the parallel rays from the distant object, they are focused at that distance from the second surface. A sphere acts like a lens focusing distant objects onto a curved image plane just outside the sphere, as shown in Figure 2.

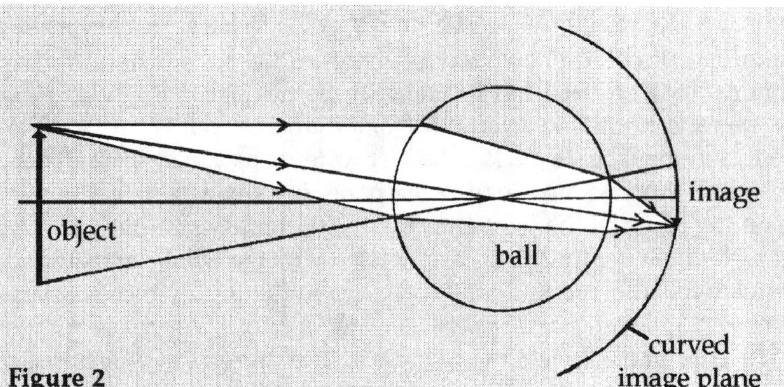

Figure 2
Image formation by a ball.

The effective index of refraction of the lens of a fish eye is typically 1.65, and the image is formed on the retina at a distance of 2.55 *R* from the center of the lens, where *R* is the radius of the lens. Amazingly, in no species of fish does that distance vary beyond the range from 2.53 *R* to 2.57 *R*. The number 2.55 is referred to as *Matthiessen's ratio.*

the center than at the surface. Most fishes focus for different distances by actually moving the lens toward or away from the retina. This is the same method of focusing as that is used in a camera. To take pictures of nearby objects, the lens of the camera is moved away from the film. The fish eye is like a water-immersed camera.

If there is no focusing at the corneal surface, its radius of curvature is of no importance. Some fish have bulbous eyes, while some have eyes with a practically flat outer surface. An eye with a flat surface could focus on objects if the eye was in air or in water.

The index of refraction of the material in the eye of the fish is, except for the lens, almost the same as that of water. As a result, the fish eye can be represented by only a spherical lens immersed in a water-like medium. The retina is a curved surface behind it, as shown in Figure 8.9b, and there is an opening in the front to admit light. The image size on the retina is found by using the rays which go through the center of the sphere. Since those rays enter and leave the sphere normal to its surface, they are not deviated, but are straight as in Figure 8.9b.

8.5 Looking From Water to Air

Another question of interest is, "What does the fish (or underwater swimmer) see when looking out from under water?" Looking toward the surface from under water, the outside world above the water appears to be concentrated

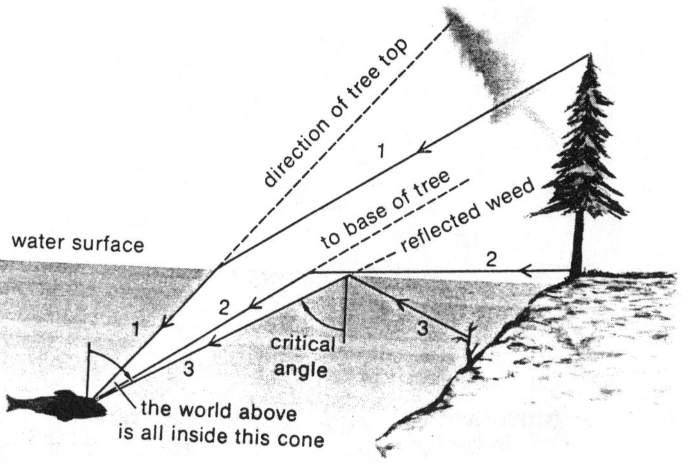

Figure 8.10
Viewing from water to air.

into a bright cone. The angle from the vertical to the edge of the cone of light is just under 50°. This cone originates because of total reflection. Figure 8.10 illustrates what happens in the various directions.

The ray marked 3 in Figure 8.10 is at the critical angle. It is only at higher angles that it is possible to see out of the water, and this is the reason for the cone of light that is seen above. This can easily be checked next time you're swimming. The circle of light on the surface will be very striking.

Box 8.2 The Ophthalmoscope

One of the earliest optical instruments to be used for medical purpose is the ophthalmoscope, invented almost a century ago. This device, which is simply a mirror with a hole in the middle, is used to examine the retina of the eye, and Figure 1 indicates how it works. Light rays from the source are reflected by a small mirror into the eye of the patient. If the eye is relaxed, the light is reflected from the retina and the rays emerge from the eye traveling in parallel lines. This parallel beam of light then passes through a small hole in the center of the mirror and into the eye of the examiner. This instrument can be used to examine the retina for abnormal conditions, such as detachment. It can also be used to give limited information about the refractive properties of the eye lens. If the lens has a problem associated with refraction, a set of lenses on a wheel attached to the mirror of the ophthalmoscope can be used to bring the retina into focus.

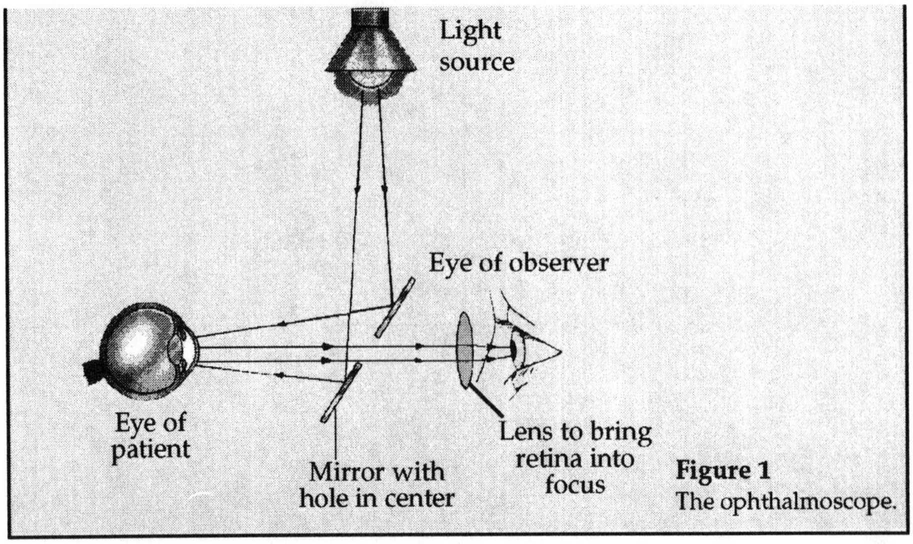

Figure 1
The ophthalmoscope.

8.6 Diffraction Effects in the Eye

A point of light on an object is focused by the eye as a diffraction spot on the retina. Two object points cannot be seen as two if the diffraction spots overlap. It is interesting to compare these diffraction spot sizes to receptor cell spacing and to see what factors in the eye affect the diffraction spot. Whereas the resolution of the human eye has been calculated on the basis of receptor cell spacing, it can also be calculated by considering diffraction; the results can then be compared. The diffraction effects are present in the eye of any creature, setting a limit to visual acuity no matter what the spacing of receptor cells.

To analyze the phenomenon, consider a wave just passing through the pupil and converging to form an image point at *P* on the retina, as in Figure 8.11. The optical paths *AP* and *BP* are equal. The image at *P* will not be a perfectly sharp point, but will drop to zero intensity at the point P' for which the optical path difference $AP' - BP'$ is 1.22 times half a wavelength (the 1.22 is the factor for a circular aperture).* The analysis is almost exactly the same as that for the camera. An important difference is that the wavelength in the medium must be considered; this is the wavelength in air (λ) divided by the index (1.336 for the eye). Letting λ_m be the wavelength in the medium,

$$\lambda_m = \frac{\lambda}{n}$$

Performing the analysis as for the camera, the radius of the diffraction spot on the retina is given from

$$\frac{1.22\ \lambda_m/2}{a/2} = \frac{r}{l}$$

* Note that not all animals have circular iris openings.

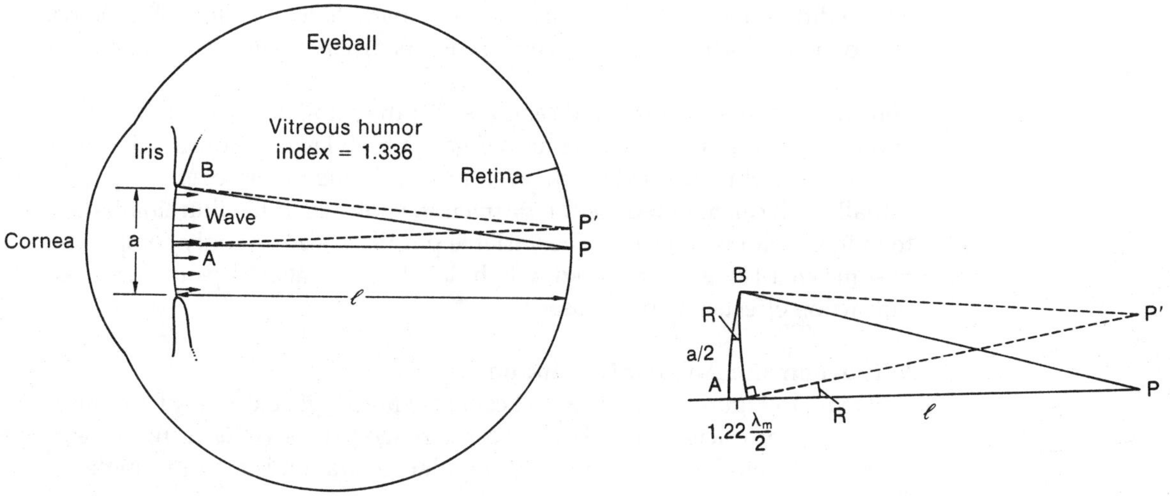

Figure 8.11
Diagrams to assist in finding the size of the diffraction spot on the retina.

Example 8.4

Find the radius of the diffraction spots in human eye for pupillary sizes of 2 mm and 5 mm. Use the wavelength in the middle of the visible spectrum (in air, 0.55 *n*m); in the eye, where *n* is 1.33, the wavelength λ_m is close to 0.4 *n*m. The length of the eyeball from iris (aperture) to retina is 20 mm.

For $a = 2$ mm,

$$r = 1.22 \times 0.4\ nm \times 20\ mm/2\ mm$$

$$= 4.9\ nm$$

For $a = 5$ mm, by a similar process,

$$r = 2.0\ nm$$

The distance *l* is from the iris to the retina. Solving for *r*, the radius of the diffraction spot on the retina, yields

$$r = 1.22\ \lambda_m \frac{l}{a} \quad \textbf{(8.5)}$$

The results of Example 8.4 are rather startling because, with the pupil closed down, the diffraction spots on the retina are much larger than the receptor cell spacing in the most sensitive part of the eye. Only with the pupil wide open do the diffraction spots shrink to a size comparable to receptor cell spacing. Acuity on the basis of receptor cell spacing is discussed in Section 8.3.

Perception of maximum detail requires that the pupil be wide open, yet we know that in order to see maximum detail, a high level of illumination must be used. In doing detailed work, even in a fairly high light level, the pupil actually will dilate to reduce the diffraction spot size. If the illumination is too high, detail will be lost. The variable pupillary size not only compensates to some small extent for changing light levels, but it also helps to minimize diffraction effects when necessary.

Diffraction and Angular Resolution

Referring to Figure 8.11, P is at the center of an image and P' is at the minimum of its diffraction spot. For a second image to be seen as such, its maximum can be as close as P' to P. The angular separation is then the angle PCP', shown as R.

The angular resolution is then, in radians, given by

$$R = \frac{r}{l}$$

Example 8.5

Calculate the possible angular resolution of eyes with pupillary diameters of 0.5 mm, 1 mm, 2 mm, 5 mm, and 10 mm. Express the answers in minutes and seconds of angle. In taking the ratio, l and a must be in the same units; so let $\lambda = 0.55\ nm = 0.55 \times 10^{-3}$ mm, and then a may be in m. Use $n = 1.33$ for an average eye, and also divide by 0.0003 or 0.3×10^{-3} to convert from radians to minutes of angle. Then for R in minutes,

$$R = (1.22 \times 0.55 \times 10^{-3}\ \text{mm})/(1.33 \times 0.3 \times 10^{-3} \times a)$$
$$= 1.7\ \text{mm}/a$$

Next tabulate values of a and 1.7 mm/a (which is R):

APERTURE ***a***, mm	RESOLVING POWER ***R***, MINUTES OR SECONDS
0.5	3.4′
1	1.7′
2	0.9′
5	0.34′ = 20″
10	0.12′ = 10″

but r is given by 1.22 $\lambda_m l/a$; substituting this into the equation, the eyeball length l cancels to leave

$$R = 1.22\ \lambda \frac{m}{a}$$

In terms of wavelength λ in air, $\lambda_m = \lambda/n$, where n is the index of the medium in the eye; the angular resolution in radians is thus

$$R = 1.22\ \lambda/na$$

This is interesting, because one factor that can vary appreciably in the eyes of different animals is the diameter of the aperature a, and the resolving power varies inversely as a. The larger the pupillary diameter, the better the possible resolution.

Visual acuity depends on two effects other than the sharpness of focusing. These are receptor cell size and diffraction effects. At the limit of vision for the human eye, these two are approximately equal. Sharper vision would not result from smaller receptor cells, because then the diffraction effects would predominate. If diffraction effects were reduced by an increased pupillary diameter, for example, no increased sharpness would result unless receptor cell spacing was reduced. These are all physical effects which apply to the eye of any creature; no other creature can have sharper vision than man unless the physical factors such as dimensions of the eye (length and pupillary size) and receptor cell spacing differ in such a way as to allow improved vision.

The hawk *Buteo buteo* is one of those birds that apparently has exceptionally keen vision. The receptor cells on the retina of that bird are packed in densities up to a million per square millimeter. This is a spacing of 1 nm between centers, which is about the same as that in the fovea of man. The pupillary diameter is not unlike that of man, also; so on the basis of the physical factors that affect sharpness of vision, that bird cannot see more distinctly than man. This raises a further question; why does the hawk *apparently* have very sharp vision?

This bird and others with apparently very keen vision can see mice or other small animals from great altitudes. But this does not involve what is called resolution. In speaking of the resolution of the eye in a case such as this, the question is whether the bird can tell from a great height if there is one mouse alone or if there are two mice standing side by side. The bird may see one mouse as a point of some sort of "color" on the ground. In seeing a single thing as a "point," the Rayleigh criterion is not relevant at all. For instance, the angular diameters of stars viewed from earth are mostly below a hundredth of a second of angle. The resolving power of our eye is, at best, 20 seconds of angle, yet we see stars. If there are two stars close together, we do not see them as two unless their angular separation is much more than 20 seconds. In fact, using our night vision, our resolution is much worse than this.

Section 9

THE ELECTRON MICROSCOPE

A practical device that relies on the wave characteristics of electrons is the electron microscope (Figure 9.1a), which is in many respects similar to an ordinary compound microscope. One important difference between the two is that the electron microscope has a much greater resolving power because electrons can be accelerated to very high kinetic energies, giving them a very short wavelength. Any microscope is capable of detecting details that are comparable in size to the wavelength of the radiation used to illuminate the object. Typically, the wavelengths of electrons are about 100 times shorter than those of the visible light used in optical microscopes. As a result, electron microscopes are able to distinguish details about 100 times smaller.

In operation, a beam of electrons falls on a thin slice of the material to be examined. The section to be examined must be very thin, typically a few hundred angstroms, in order to minimize undesirable effects, such as absorption or scattering of the electrons. The electron beam is controlled by electrostatic or magnetic deflection, which acts on the charges to focus the beam to an image. Rather than examining the image through an eyepiece as in an ordinary microscope, a magnetic lens forms an image on a fluorescent screen. the fluorescent screen is necessary because the image produced would not otherwise be visible. An example of a photograph taken by an electron microscope is shown in Figure 9.1b.

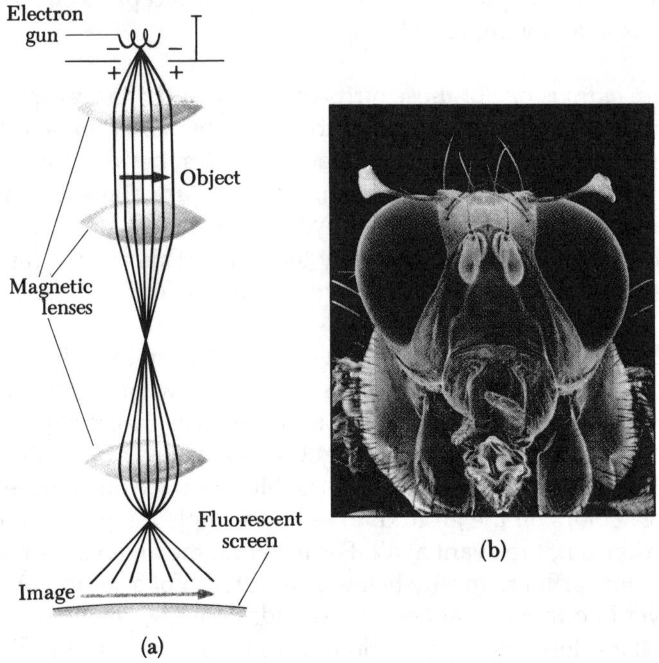

Figure 9.1
(a) Diagram of an electron microscope. The "lenses" that control the electron beam are magnetic deflection coils.
(b) A scanning electron micrograph (SEM) of a Mediterranean fruitfly. (© *David Scharf/Peter Arnold, Inc.*)

Section 10
MODERN PHYSICS

10.1 Radioactivity

In 1896 Becquerel discovered that an ore containing uranium emits an invisible radiation that can expose a photographic plate. He called this unknown type of radiation x-rays, and soon the substances that exhibited this phenomenon were said to be **radioactive**. There are three types of radiation that can be emitted by a radioactive substance: alpha (α) decay, in which the emitted particles are helium nuclei; beta (β) decay, in which the emitted rays are electrons; and gamma (γ) decay, in which the emitted rays are high energy photons.

It can be shown that the number of nuclei present varies with time exponentially according to the expression

$$N = N_0 e^{-\lambda t}. \quad \textbf{(10.1)}$$

where N is the number of radioactive nuclei present at time t, N_0 is the number present at time $t = 0$, and λ is a constant called the **decay constant**. In fact, any quantity proportional to the number of nuclei present obeys a similar decay equation. As examples, the amount of mass, m, present or the decay rate, R, also exponentially decay according to equation 10.1.

One of the most common terms encountered in any discussion of radioactive materials is **half-life,** $T_{1/2}$. This is because all radioactive substances follow the same general decay pattern. After a certain interval of time, half of the original number of nuclei present in a sample will have decayed; then in a second time interval equal to the first, half of those nuclei remaining will have decayed, and so on. The half-life and the decay constant are related as

$$T_{1/2} = \frac{0.693}{\lambda} \quad \textbf{(10.2)}$$

10.2 Carbon Dating and Radon Detecting

Carbon Dating

The beta decay of ^{14}C is commonly used to date organic samples. Cosmic rays (high-energy particles from outer space) in the upper atmosphere cause nuclear reactions that create ^{14}C from ^{14}N. In fact, the ratio of ^{14}C to ^{12}C isotope abundance in the carbon dioxide molecules of our atmosphere has a constant value of about 1.3×10^{-12} as determined by measuring carbon ratios in tree rings. All living organisms have the same ratio of ^{14}C to ^{12}C because they continuously exchange carbon dioxide with their surroundings. When an organism dies, however, it no longer absorbs ^{14}C from the atmosphere, and so the ratio of ^{14}C to ^{12}C decreases as the result of the beta decay of ^{14}C. It is therefore possible to measure the age of a material by measuring its activity per unit mass due to the decay of ^{14}C. Using carbon dating, samples of wood, charcoal, bone, and shell have been identified as having lived from 1000 to 25,000 years ago. This knowledge has helped us to reconstruct the history of living organisms—including humans—during this time span.

A particularly interesting example is the dating of the Dead Sea Scrolls. This group of manuscripts was first discovered by a shepherd in 1947. Translation showed them to be religious documents, including most of the books of the Old Testament. Because of their historical and religious significance, scholars wanted to know their age. Carbon dating applied to fragments of the scrolls and to the material in which they were wrapped established their age at about 1950 years.

Radon Detecting

Radioactivity can also affect our daily lives in harmful ways. Soon after the discovery of radium by the Curies, it was found that the air in contact with radium compounds becomes radioactive. It was shown that this radioactivity came from the radium itself, and the product was therefore called "radium emanation." Rutherford and Soddy succeeded in condensing this "emanation," confirming that it is a real substance - the inert, gaseous element now called radon, Rn. We now know that the air in uranium mines is radioactive because of the presence of radon gas. The mines must therefore be well ventilated to help protect the miners. The fear of radon pollution has now moved from uranium mines into our own homes. Since certain types of rock, soil, brick, and concrete contain small quantities of radium, some of the resulting radon gas finds its way into our homes and other buildings. The most serious problems arise from leakage of radon from the ground into the structure. One practical remedy is to exhaust the air through a pipe just above the underlying soil or gravel directly to the outdoors by means of a small fan or blower.

Example 10.1 Should We Report This to Homicide?

A 50-g sample of carbon is taken from the pelvis bone of a skeleton and is found to have a carbon-14 decay rate of 200 decays/min. It is known that carbon from a living organism has a decay rate of 15 decays/min · g and that ^{14}C has a half-life of 5730 y= 3.01×10^9 min. Find the age of the skeleton.

Solution

Let us start with

$$R = R_0 e^{-\lambda t}$$

where R is the present activity, and R_0 was the activity when the skeleton was a part of a living organism. We are given that $R = 200$ decays/min, and we can find R_0 as

$$R_0 = \left(15 \frac{\text{decays}}{\text{min} \cdot \text{g}}\right)(50\ \text{g}) = 750 \frac{\text{decays}}{\text{min}}$$

The decay constant is found from Equation 10.2 as

$$\lambda = \frac{0.693}{T_{1/2}} = \frac{0.693}{3.01 \times 10^9\ \text{min}} = 2.30 \times 10^{-10}\ \text{min}^{-1}$$

Thus, we make the following substitutions:

$$R = R_0 e^{-\lambda t}$$

$$200 \frac{\text{decays}}{\text{min}} = \left(750 \frac{\text{decays}}{\text{min}}\right) e^{-(2.30 \times 10^{-10}\ \text{min}^{-1})t}$$

or

$$0.266 = e^{-(2.30 \times 10^{-10}\ \text{min}^{-1})t}$$

Now, we take the natural log of both sides of the equation, to give

$$\ln(0.266) = -(2.30 \times 10^{-10}\ \text{min}^{-1})t$$

$$-1.32 = -(2.30 \times 10^{-10}\ \text{min}^{-1})t$$

$$t = 5.74 \times 10^9\ \text{min} = 10{,}930\ \text{y}$$

Carbon-14 and the Shroud of Turin

Since the Middle Ages, many people have marveled at a 14-foot long, yellowing piece of linen found in Turin, Italy, which is purported to be the burial shroud of Jesus Christ. The cloth bears a remarkable, full-size likeness of a crucified body, with wounds that could have been caused by a crown of thorns and a wound in the side that could have been the cause of death. The skepticism over the authenticity of the shroud has existed since its first public showing in 1354; in fact, a French bishop declared it to be a fraud at the time. Because of its controversial nature, religious bodies have taken a neutral stance on its authenticity.

In 1978, the bishop of Turin allowed the cloth to be subjected to scientific analysis, but notably missing from these tests was a carbon-14 dating. The reason for this omission was that, at the time, carbon dating techniques required a piece of cloth about the size of a handkerchief. In 1988, the process had been refined to the point that pieces as small as one inch on a side would be sufficient, and permission was granted to allow the dating to proceed. Three labs were selected for the testing, and each was given four pieces of material. One of these was a piece of the shroud while the other three were control pieces similar in appearance to the shroud.

The testing procedure consisted of burning the cloth to produce carbon dioxide, which was then converted chemically to graphite, and the graphite sample was subjected to carbon-14 analysis. The end result was that all three labs agreed amazingly well on the age of the shroud. The average of their results gave a date for the cloth of A.D. 1320 $\pm$ 60 years, with an absolute assurance that the cloth is no older than A.D. 1200. Thus, carbon-14 dating has unraveled the most important mystery concerning the shroud, but others remain. For example, the investigators have not as yet been able to explain how the image was imprinted.

Three Mile Island and Chernobyl

The two most publicized nuclear accidents have been those at the Three Mile Island nuclear plant in Middletown, Pennsylvania, and at the Soviet reactor at Chernobyl, Russia. Let us describe the events and consequences of these accidents.

The series of events that caused the Three Mile Island incident began at 4 A.M. on March 28, 1979, with a routine maintenance operation. Workers changing the water purifier in a part of the piping system at the plant inadvertently allowed some air into the pipe, which caused an interruption in the flow of water to the reactor. If such an event occurs, backup systems are installed to respond automatically, but in this case several things went wrong. Two pumps in the backup system were out of service for routine maintenance and, therefore, could not deliver water for cooling. Reactor operators did not notice the water-pump problem because warning lights were obscured with tags that had been placed on the pumps when they were taken out of service. The loss of cooling water to the reactor caused a pressure build-up in the reactor core. This caused a relief valve in the primary coolant loop to open automatically and release superheated steam. This automatic sequence led to further difficulties because the valve failed to close after venting, and a dangerous drop in pressure occurred inside the loop. At this point, the emergency cooling system was activated and pressure gauges in the control room indicated that the water level was being maintained such that the fuel rods were covered. This, however, was a false reading because a 1000 ft^3 bubble of steam and hydrogen gas was pushing the water level up, leaving part of the core exposed. This caused the core to overheat to the point where temperatures inside the reactor vessel were so high that they could not be measured on instruments at the plant. When the situation was finally brought under control about three days later, it was found that the fuel rods had cracked, leading to contamination of the pressure vessel. The cleanup was a major problem and continues to be difficult to handle. Fortunately, there was no major release of radiation to the public. The radiation release was such that the public received no more than about 10% of what would be normally received from natural sources in one year. The subsequent threat to public health via cancers, if any, is too low to estimate or detect in any reliable way.

A far more serious accident occurred at the Soviet reactor at Chernobyl, Russia, in 1986. This disaster started when the cooling system failed, which caused an overheating and ultimate melting of the uranium oxide fuel rods. The next stage in the disaster occurred when the operators were late in flooding the core with water. The already hot core changed the water to steam, which in turn reacted with the graphite moderator in the core, the melted uranium, and other molten substances to produce a mixture of hydrogen, oxygen, methane, and carbon monoxide. These gases then interacted violently to produce a chemical explosion, which ruptured the walls of the plant. The plant was built without the added protection of a containment vessel, and enormous levels of radioactive materials were spewed into the atmosphere for several days.

Heroic measures were required to bring the fire and release of radioisotopes into the environment under control. Several tons of sand, clay, lead, and boron were dropped on the reactor from above by helicopters, and other workers tunneled under the plant to seal it from below with concrete, lead, and boron in an unsuccessful attempt to isolate the radioactive materials from the groundwater. Virtually all of these workers, and pilots have now died from leukemia induced from the radiation dose received. At present an area of about 700 km ($250 mi^2$) is so contaminated that it is a wasteland and will be so for decades. The final story of the death and devastation caused by the Chernobyl incident is yet to be written. It should be noted here that the Soviet reactor was old and out of date and lacked most of the safety features, such as a containment vessel, inherent to U.S. designs. A disaster of these proportions is very unlikely to happen in a properly designed power plant.

10.3 Radioactive Tracing and Activation Analysis

Tracing

Radioactive particles can be used to trace chemicals participating in various reactions. One of the most valuable uses of radioactive tracers is in medicine. For example, ^{131}I is an artificially produced isotope of iodine (the natural, nonradioactive isotope is ^{127}I). Iodine, which is a necessary nutrient for our bodies, is obtained largely through the intake of iodized salt and seafood. The thyroid gland plays a major role in the distribution of iodine throughout the body. In

order to evaluate the performance of the thyroid, the patient drinks a very small amount of radioactive sodium iodide. Two hours later, the amount of iodine in

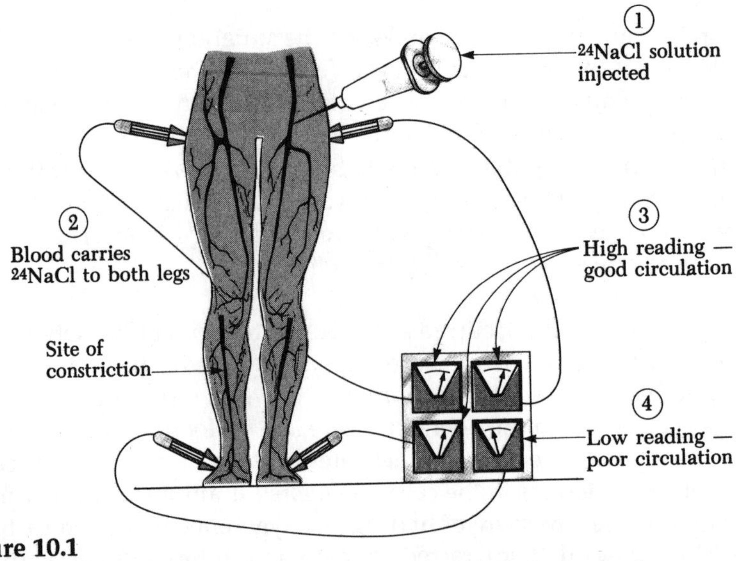

Figure 10.1
A tracer technique for determing the condition of the human circulatory system.

the thyroid gland is determined by measuring the radiation intensity at the neck area.

A second medical application is indicated in Figure 10.1. Here a salt containing radioactive sodium is injected into a vein in the leg. The time at which the radioisotope arrives at another part of the body is detected with a radiation counter. The elapsed time is a good indication of the presence or absence of constrictions in the circulatory system.

The tracer technique is also useful in agricultural research. Suppose the best method of fertilizing a plant is to be determined. A certain material in the fertilizer, such as nitrogens, can be tagged with one of its radioactive isotopes. The fertilizer is then sprayed on one group of plants, sprinkled on the ground for a second group, and raked into the soil for a third. A Geiger counter is then used to track the nitrogen through the three types of plants.

Tracing techniques are as wide-ranging as human ingenuity can devise. Present applications range from checking the absorption of fluorine by teeth to checking contamination of food-processing equipment by cleansers to monitoring deterioration inside an automobile engine. In the latter case, a radioactive material is used in the manufacture of the pistons, and the oil is checked for radioactivity to determine the amount of wear on the pistons.

Activation Analysis

For centuries, a standard method of identifying the elements in a sample of material has been chemical analysis, which involves testing a portion of the material for reactions with various chemicals. A second method is spectral analysis, which utilizes the fact that, when excited, each element emits its own characteristic set of electromagnetic wavelengths. These methods are now supplemented by a third technique, **neutron activation analysis.** Both chemical and spectral methods have the disadvantage that a fairly large sample of the material must be destroyed for the analysis. In addition, extremely small quantities of an element may go undetected by either method. Activation analysis has an advantage over the other two methods in both of these respects.

When the material under investigation is irradiated with neutrons, nuclei in the material will absorb the neutrons and be changed to different isotopes. Most of these isotopes will be radioactive. For example, ^{65}Cu absorbs a neutron to become ^{66}Cu, which undergoes beta decay:

$$^{1}_{0}n + ^{65}_{29}Cu \rightarrow ^{66}_{30}Cu \rightarrow Zn + ^{0}_{-1}e \qquad \textbf{(10.3)}$$

The presence of the copper can be deduced because it is known that ^{66}Cu has a half-life of 5.1 min and decays with the emission of beta particles having maximum energies of 2.63 and 1.59 MeV. Also emitted in the decay of 66_{Zn} is a gamma ray having an energy of 1.04 MeV. Thus, by examining the radiation emitted by a substance after it has been exposed to neutron irradiation, one can detect extremely small traces of an element.

Neutron activation analysis is used routinely by a number of industries, but the following nonroutine example of its use is of interest. Napoleon died on the island of St. Helena in 1821, supposedly of natural causes. Over the years, suspicion has existed that his death was not all that natural. After his death, his head was shaved and locks of his hair were sold as souvenirs. In 1961, the amount of arsenic in a sample of this hair was measured by neutron activation analysis. Unusually large quantities of arsenic were found in the hair. (Activation analysis is so sensitive that very small pieces of a single hair could be analyzed.) Results showed that the arsenic was fed to him irregularly. In fact, the arsenic concentration pattern corresponded to the fluctuations in the severity of Napoleon's illness as determined from historical records.

10.4 Medical Imaging Techniques: The CAT Scan and MRI

Computed Axial Tomography (CAT Scans)

The normal x-ray of a human body has two primary disadvantages when used as a source of clinical diagnosis. First, it is difficult to distinguish between various types of tissue in the body because they all have similar x-ray absorption properties. Second, a conventional x-ray absorption picture is indicative of the average amount of absorption along a particular direction in the body, leading to somewhat obscured pictures. To overcome these problems, a device called a CAT scanner was developed in England in 1973; it is capable of producing pictures of much greater clarity and detail than were previously obtainable.

The operation of a CAT scanner can be understood by considering the following hypothetical experiment. Suppose a box consists of four compartments, labeled A,B,C, and D as in Figure 10.2a. Each compartment has a different amount of absorbing material than any other compartment. What set of experimental procedures will enable us to determine the relative amounts of material in each compartment? The following steps outline one method that will provide this information. First, a beam of x-rays is passed through compartments A and C, as in Figure 10.2b. The intensity of the exiting radiation is reduced by absorption by some number that we assign as 8. (The number 8 could mean, for example, that the intensity of the exiting beam is reduced by eight-tenths of one percent from its initial value.) Since we do not know which of the compartments, A or C, was responsible for this reduction in intensity, half the loss is assigned to each compartment as in Figure 10.2c. Next, a beam of x-rays is passed through compartments B and D, as in Figure 10.2b. The reduction in intensity for this beam is 10, and again we assign half the loss to each compartment. We now redirect the x-ray source so that it sends one beam through compartments A and B and another through compartments C and D, as in Figure 10.2d, and again measure the absorption. Suppose the absorption through compartments A and B in this experiment is measured to be 7 units. On the basis of our first experiment, we would have guessed it would be 9 units, 4 by compartment A and 5 by compartment B.

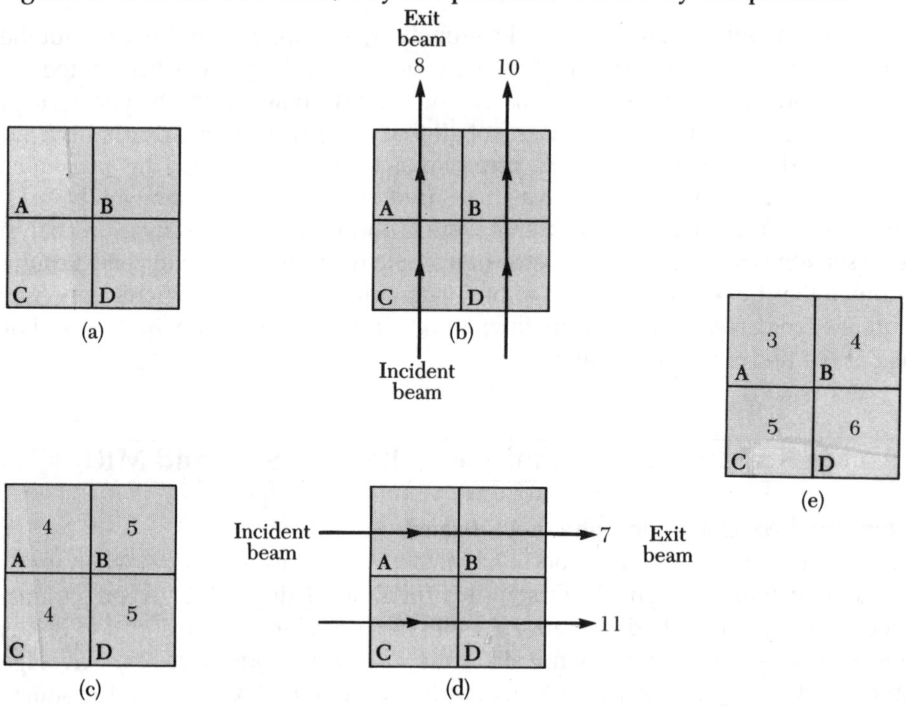

Figure 10.2
An experimental procedure for determining the relative amounts of x-ray absorption by four different compartments in a box.

Thus, we have reduced the guessed absorption for each compartment by 1 unit so that the sum is 7 rather than 9, to give the numbers shown in Figure 10.2e. Likewise, when the beam is passed through compartments C and D as in Figure 10.2d, we may find the total absorption to be 11 as compared to our first experiment of 9. In this case, we add 1 unit of absorption to each compartment to give a sum of 11 as in Figure 10.2e. This somewhat crude procedure could be improved by measuring the absorption along other paths. However, these simple measurements are sufficient to enable us to conclude that compartment D contains the most absorbing material and A the least. A visual representation of these results can be obtained by assigning to each compartment a shade of gray corresponding to the particular number associated with the absorption. In our example, compartment D would be very dark while compartment A would be very light.

The steps outlined above are representative of how a CAT scanner produces images of the human body. A thin slice of the body is subdivided into perhaps 10,000 compartments , rather than 4 compartments as in our simple example. The function of the CAT scanner is to determine the relative absorption in each of these 10,000 compartments and to display a picture of its calculations in various shades of gray. Note that CAT stands for **computed axial tomography**. The term axial is used because the slice of the body to be analyzed corresponds to a plane perpendicular to the head-to-toe axis. *Tomos* is the Greek word for slice and *graph* is the Greek word for picture. In a typical diagnosis, the patient is placed in the position shown in Figure 10.3 and a narrow beam of x-rays is sent through the plane of interest. The emerging x-rays are detected and measured by photomultiplier tubes behind the patient. The x-ray tube is then rotated a few degrees, and the intensity is recorded again. An extensive amount of information is obtained by rotating the beam through 180 degrees at intervals of about one degree per measurement, resulting in a set of numbers assigned to each of the 10,000 "compartments" in the slice. These numbers are then converted by the computer to a photograph in various shades of gray for this segment of the body.

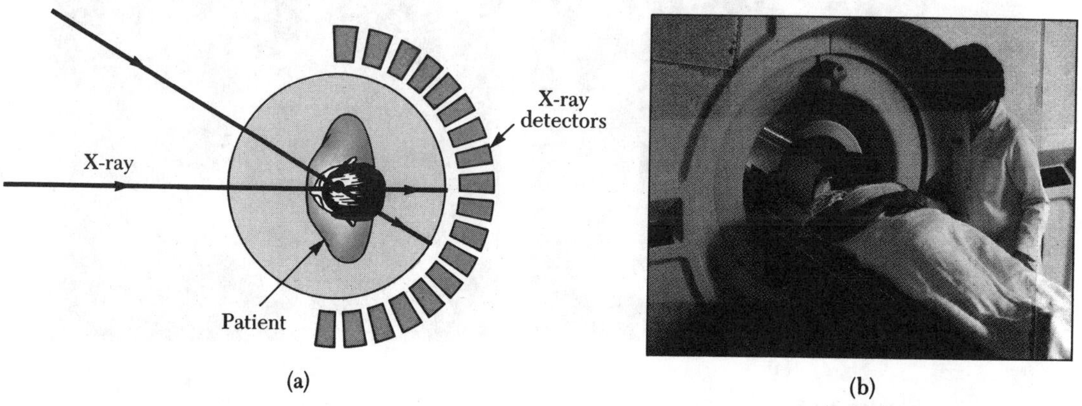

Figure 10.3
(a) **CAT** scanner detector assembly. (b) Photograph of a patient undergoing a CAT scan in a hospital. (*Jay Freis/The IMAGE Baank*)

A brain scan of a patient can now be made in about 2 seconds, while a full-body scan requires about 6 seconds. The final result is a picture containing much greater quantitative information and clarity than a conventional x-ray photograph. Since **CAT** scanners use x-rays, which are an ionizing form of radiation, the technique presents a health risk to the patient being diagnosed.

Magnetic Resonance Imaging (MRI)

At the heart of magnetic resonance imaging (MRI) is the fact that when a nucleus having a magnetic moment is placed in an external magnetic field, its moment will precess about the magnetic field with a frequency that is proportional to the field. For example, a proton, whose spin is 1/2, can occupy one of two energy states when placed in an external magnetic field. The lower energy state corresponds to the case where the spin is aligned with the field, while the higher energy state corresponds to the case where the spin is opposite the field. Transitions between these two states can be observed using a technique known as nuclear magnetic resonance. A dc magnetic field is applied to align the magnetic moments, while a second, weak oscillating magnetic field is applied perpendicular to the dc field. When the frequency of the oscillating field is adjusted to match the precessional frequency of the magnetic moments, the nuclei will "flip" between the two spin states. These transitions result in a net absorption of energy by the spin system, which can be detected electronically.

In MRI, image reconstruction is obtained using spatially varying magnetic fields and a procedure for encoding each point in the sample being imaged. Some MRI images taken on a human head are shown in Figure 10.4. In practice, a computer-controlled pulse sequencing technique is used to produce signals that are captured by a suitable processing device. This signal is then subjected to appropriate mathematical manipulations to provide data for the final image. The main advantage of MRI over other imaging techniques in

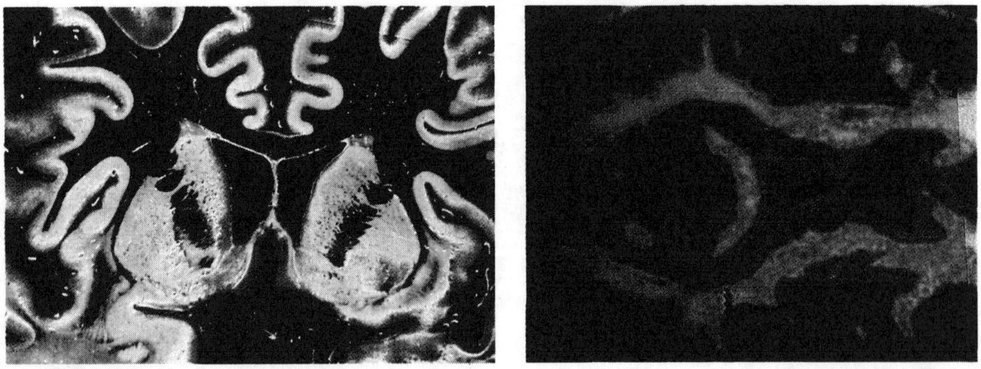

(a) (b)

Figure 10.4

State-of-the-art magnetic resonance images taken of the human head. (a) is a sagittal view and (b) is a coronal view. The slice images are of 1 cm thickness and the result of a 128 ×128 point reconstruction (*Photo Researchers, Inc.*)

medical diagnostics is that it causes minimal damage to cellular structures. Photons associated with the rf signals used in MRI have energies of only about 10^{-7}eV. Since molecular bond strengths are much larger (of the order of 1 eV), the rf photons cause little cellular damage. In comparison, x-rays or γ-rays have energies ranging from 10^4 to 10^6eV and can cause considerable cellular damage.